A portion of the p  ..... oook will go to
A Child is Missing

The phone rang, and a man's voice said, "This is George LaMoureaux, I am calling from Anchorage, Alaska. I heard about your program to assist in finding missing children and I want to help you. I would like you to come to Anchorage and start the program in Alaska."

To say the least, I was stunned. Not many people stepped up to the plate to help bring this program to fruition anywhere, only a handful. We did not have a lot of funding, in fact they were my funds and they were lean. I asked George if he could help with funding and he did in one way or the other.

As I read *Everest—A Triumph in Adversity*, I could not put it down. Being a slow reader, I finished the book in two days. Learning more about George led me to believe he was on the right track with his life and the gifts he was given to accomplish the amazing things he has performed.

I know how it feels… when I got the message to find missing children, the elderly, and the challenged, I knew I had a purpose for my life. People ask me why I started A Child Is Missing and I just reply, "It came from above. That's all I know!" They always look at me kind of strange. I believe that you are where you are when you are there, and you are destined to be there. Believe.

George never gave up on A Child Is Missing and continues to let people know there is a way to assist law enforcement find missing children. George and others made the Climb for

America's Children up Denali, then Everest because he and they believed the program would work to save lives and it has.

The use of technology for these purposes was not thought about 20 years ago. We now have assisted law enforcement in the safe recovery of over 2500 children! Technology plays a huge role in our lives in many different areas. And today, we understand more fully how technology can help us locate and save missing children. Thankfully, with George's help, America has a way to find our missing children.

(For more information or to donate to A Child Is Missing, please go to: *www.achildismissing.org.*)

—*Sherry Friedlander, Founder A Child Is Missing*

EVEREST
— A —
TRIUMPH IN ADVERSITY

EVEREST
— A —
Triumph in Adversity

*A True Story of Faith
in the Face of Extreme Adversity*

Entrepreneur, Cancer Survivor, Philanthropist,
McKinley & Everest Mountaineer

George LaMoureaux

With Rob Fischer
& Foreword by Dr. Jerry Prevo

EVEREST — A TRIUMPH IN ADVERSITY

Copyright ©2018 George LaMoureaux

ISBN-13: 978-1984220523

THE HOLY BIBLE, NEW INTERNATIONAL VERSION®, NIV® Copyright © 1973, 1978, 1984, 2011 by Biblica, Inc.® Used by permission. All rights reserved worldwide.

The ESV® Bible (The Holy Bible, English Standard Version®) copyright © 2001 by Crossway, a publishing ministry of Good News Publishers. ESV® Text Edition: 2011. The ESV® text has been reproduced in cooperation with and by permission of Good News Publishers. Unauthorized reproduction of this publication is prohibited.
All rights reserved.

Scripture taken from the NEW AMERICAN STANDARD BIBLE®, Copyright © 1960,1 962,1963,1968,1971,1972,1973,1975,1977,1995 by The Lockman Foundation.
Used by permission.

Holy Bible, New Living Translation copyright © 1996, 2004, 2007 by Tyndale House Foundation. Used by permission of Tyndale House Publishers Inc., Carol Stream, Illinois 60188. All rights reserved. New Living, NLT, and the New Living Translation logo are registered trademarks of Tyndale House Publishers.

DEDICATION

I dedicate this book to my family whom I love so very much. They have stood with me my entire life.

Furthermore, I dedicate this book to my friends who, through thick and thin, have continued to be my friends.

Finally, I dedicate this book to all of you suffering in your day-to-day existence who need to be inspired and empowered to "never quit." When you're hardest hit with debilitating diseases like cancer, or loss of family, friends, relationships or finances, I encourage you to keep going in the face of extreme adversity. Know that as long as you have your hope based in your relationship with Jesus Christ no matter what happens, you'll be blessed!

ACKNOWLEDGEMENTS

As you read this book, you'll no doubt recognize how important family is to me. In that spirit, I am so grateful to my family for standing with me and loving me all these years through both good and difficult times. I love you all!

Also, I'm very thankful to my pastor, Dr. Jerry Prevo, for reading my book while in manuscript form and for writing the Foreword. I'd like to also express my gratitude to Dr. Ken Friendly, the author of *Never Quit* (included at the end of this volume). His message contained in *Never Quit* inspired and motivated me to keep on going in spite of incredible odds.

I wish to thank the many friends and people mentioned in this book and numerous others not mentioned in this book who have invested in my life and business and who have contributed to my success. I don't want anyone left out. Finally, I'm grateful, for Rob Fischer, without whom I would not have been able to complete this project.

—George LaMoureaux,
Anchorage, Alaska

TABLE OF CONTENTS

Dedication .... *v*
Acknowledgements .... *vii*
Foreword .... *xiii*
Introduction .... *xv*

Chapter One — The Early Years 1

Chapter Two — A Butt-Whoopin' 9

Chapter Three — Tougher than
Boards and Bricks 17

Chapter Four — A Pioneer, Not a Farmer 25

Chapter Five — "Chicken George" 31

Chapter Six — Wallowing with the Pigs 37

Chapter Seven — From High Life to No Life 43

Chapter Eight — She Loved Cartoons 49

Chapter Nine — Laughter and Adventure Coming
Your Way! 53

Chapter Ten — KAPTYH —(Cartoon in Russian) 59

Chapter Eleven — Riding the Wave 67

Chapter Twelve — From Hero to Zero 73

Chapter Thirteen — Selling Ice Water to Eskimos 83

Chapter Fourteen — Mergers, Acquisitions, and Other
  Developments 91

Chapter Fifteen — The Resorts 99

Chapter Sixteen — A New Life 113

Chapter Seventeen — A Lofty Vision 119

Chapter Eighteen — The Coldest Mountain on Earth. 125

Chapter Nineteen — "The High One" 131

Chapter Twenty — Keep Swinging for the Fence! 141

Chapter Twenty-One — Joys and Sadness,
  the Stuff of Life 149

Chapter Twenty-Two — The Odds Are Good,
  but the Goods Are Odd 159

Chapter Twenty-Three — Cancer 165

Chapter Twenty-Four — Never Quit! 173

Chapter Twenty-Five — Mt. Everest 183

Chapter Twenty-Six — Waiting on the Chinese 191

Chapter Twenty-Seven — The Summit! 199

Chapter Twenty-Eight — Distress Call 207

Chapter Twenty-Nine — Gerson Therapy 227

Chapter Thirty — Shoot for the Moon! 235

Chapter Thirty-One — Buying the World's Largest
Library of Motion Pictures (MGM)! 245

Chapter Thirty-Two — I Felt Like Job! 253

Chapter Thirty-Three — The Tallest Cross 259

Chapter Thirty-Four — From Cross to Crosshairs 267

Chapter Thirty-Five — Thinking Big! 271

Chapter Thirty Six — Not a Spectator 281

Chapter Thirty-Seven — Be a Finisher! 295

Appendix: Never Quit! by Dr. Kenneth Friendly 305

Editor's Note 381

Recommended Reading 385

FOREWORD

As you read the story of George LaMoureaux, you will be amazed at his ability to come up with great ideas. You will be amazed at the people he has come in contact with and worked with.

George's life has been like a roller coaster with many highs, lows, twists and turns—so many lows that you would expect him to be depressed, discouraged and disillusioned. But he is still determined!

He is determined to complete his life committed to his faith, family and friends. Among the mother lodes that George has hit financially, he has cashed in on the mother lode of mother lodes through a sincere relationship with Jesus Christ.

Every Sunday George is in the front row of church taking in the preaching of God's Word with his daughter Ashley by his side. His mother was there with them until she went home to be with the Lord.

The life of George LaMoureaux is more interesting than a mystery. The mystery that remains is what "big deal" will finally materialize for a man who will not quit.

— *Dr. Jerry Prevo*

INTRODUCTION

I NEVER INTENDED to write a book, much less an autobiography. For one thing, I always thought of biographies and their like being written posthumously and I'm not dead yet! Secondly, because my life isn't over, I still have much to accomplish. My story isn't finished, so why write now?

However, as I've shared my stories with others, they've urged me to write. But their prompting alone wasn't motivation enough; I needed a loftier purpose for writing. Well, I found that noble purpose (three actually) and here we are.

First, as you read about my business pursuits and failures, my battle with cancer, the tragic loss of numerous family members, climbing Mt. McKinley and Mt. Everest, and other extreme hardships I've overcome, I pray that you will find strength to persevere through your most difficult trials and darkest hours. Whatever your personal challenges and hardships, I hope that my story will encourage and inspire you.

I am a movie fanatic. In addition to their entertainment value, many great films also offer inspiration as they portray men and women who have overcome tremendous adversity. One of my all-time favorite movies is the 1955 Audie Murphy movie *To Hell and Back*. Audie Murphy, who plays himself in this film, was the most decorated American soldier of World War II. Murphy was a true war hero.

I aspire to be like Audie Murphy. My original intent was to name my book *To Hell and Back* after Murphy's film, because the title aptly describes my life experiences. Murphy triumphed in extreme adversity and I can relate to Murphy because of the adversities I have endured.

I don't mean to put my sufferings on par with the wartime horrors that Audie Murphy and countless other soldiers have suffered. Many people endure much more difficult hardships than I have. I truly consider myself fortunate and I am grateful to be where I am and to have what I have. In some way or another, we all bear one or more severe trials. As Dr. Ken Friendly says regarding the storms of life, "Either we're in one; or we just came out of one; or there's one out there with our name on it."[1]

For some reason, seeing our own sorrows in light of others' sufferings humbles us and enables us to see our own situation more objectively. We read a story or see a movie like *To Hell and Back* and we get charged up to persevere, endure hardship, and accomplish great things. My hope is that my life's story will do that for you.

Second, as you will soon see, I've been given the gift of a few extra years of life for a reason. I don't want to squander the gifts and abilities God has given me, but to serve others and glorify Him through my talents. I've found a cause to give my life to and leave a legacy. I want to leave this world having given more than I've taken.

Third, much too late in life I came into relationship with God through Jesus Christ. Many of my pre-Christ days were filled with choices and a lifestyle that I'm not proud of today. I squandered much and sinned much, but Christ has forgiven

---

1   Dr. Kenneth Friendly, *Never Quit*, CD series (Anchorage, AK: Lighthouse Christian Fellowship, 2007).

me and given me a fresh start. I will write about my past, but I no longer live there.

So I pray that in telling my story, others will come to know Christ and His power to change their lives. And may Christ be glorified as a result.

Therefore, I have written this book to inspire and empower others not to give up when they're hardest hit, but to persevere through the toughest times, knowing that God has a plan for them. "I can do all things through Christ who strengthens me." (Philippians 4:13 author's paraphrase)

Finally, as you read my story, please remember that my life took 59 years in the making. What I mean is that this autobiography barely scratches the surface of all I have lived and experienced. There are so many other businesses, deals, relationships and other experiences that I have not detailed here. To tell all that would take me about – 59 years!

After a long day on one of my extended overseas business trips, I sat alone in a hotel processing my day. Then, as my thoughts shifted to reflect over my life, I wrote the following:

The world loves the underdogs, the dreamers out there. The world needs underdogs in order to believe that one day, maybe they too can achieve the impossible.

But, the sad truth is, underdogs seldom win.

Even so, some of the greatest achievements in all of history have been accomplished by these underdogs; these people with faith, who didn't know they couldn't do it. They took on a challenge even after suffering great personal loss time and again. These underdogs persevered with courage in the face of extreme adversity, through failure after failure. Their secret? They never quit and triumphed against all odds.

I am an underdog and this is my story.

— *George LaMoureaux,*
Anchorage, Alaska, 2015

Chapter One
— THE EARLY YEARS —

You don't have to see the whole staircase, just the first step.
Dr. Martin Luther King, Jr.

"Doc, you're holding out on me. I want you to be straight with me."

"George, the biopsy confirmed squamous-cell carcinoma of the head and neck. You're already in stage four. This is a very aggressive and deadly cancer." After an extended amount of medical detail, he summarized with, "You have perhaps 18 months to live. I'm sorry."

That was the conversation I had with my ear, nose and throat specialist nearly seven years ago. Without question, it was a conversation that has profoundly impacted my life. I'll come back to this later, but for now I'm getting ahead of myself. Let me go back to the beginning.

I was born George Austin Plein on September 14, 1956 in Oakland (Alameda), California. As one who loves research, I conducted a web search eager to discover what other monumental events accompanied my arrival into the world on that day. It's as though these great events whose birthdates we share, add significance to who we are.

However, the only other noteworthy event that occurred the day I was born was the first prefrontal lobotomy and it was performed in Washington DC! Both from my perspective and probably that of the patient, I think this moves my birth into the position of preeminence on that day. I could also comment on the significance of the fact that Washington DC is the birthplace of the lobotomy, but I'll leave that to the reader to ponder. However, it makes me wonder about the need for additional lobotomies in Washington DC with the insanity going on in our current political environment!

Other significant events that occurred in 1956 include: the release of Elvis Presley's first hit, *Heartbreak Hotel*; the Federal-Aid Highway Act initiated construction of 41,000 miles of interstate highways across the US; and the US Supreme Court declared the Alabama bus segregation laws illegal. Some of the movies to hit the big screen that year were: *The Ten Commandments* (one of my all-time favorites), *The King and I, Trapeze,* and *Around the World in Eighty Days.* The most popular musicians in 1956 included: Elvis Presley, Bill Haley and the Comets, Chuck Berry, Jerry Lee Lewis, Johnny Cash, Ella Fitzgerald, and Dean Martin.

Of course, being the first-born, September 14, 1956 was a very special day for my parents, George and Mia Plein. I know little about my father's background other than he was born in Missouri on April 23, 1924 and his mother's maiden name was Lockhart. I know, however, that my mom's veins coursed with royal blood. Her mother's maiden name was McLeod a direct descendent of the MacLeod Clan on the Isle of Skye in Scotland. (The *MacLeod* name was simplified to *McLeod* when members of the Clan immigrated to America.)

The MacLeod's are also related to the Stewarts, the current occupants of the throne of England. Grandma also traced her lineage back to Óláfr Guðrøðarson, known as King Olaf the

Black, a Norseman who ruled the Isle of Mann and part of the Hebrides. Grandma's full name reflects her noble Scottish roots: Maude Leone McLeod.

In view of my family heritage, a noteworthy event occurred on the 14th of August 1956 just a month prior to my birth. Following a visit by Her Majesty Queen Elizabeth II of England, delegates of Clan MacLeod took part in a banquet in Dunvegan Castle to celebrate the coming-of-age of the present Chief.[2]

The MacLeod Clan's motto is "Hold fast." And clearly that's what its Chiefs have done as the MacLeod Clan has owned and occupied the Dunvegan Castle for nearly eight centuries. Dunvegan Castle stands as perhaps the most prominent and historic of the family houses of Scotland. The producers of the 1986 classic movie *Highlander* starring Sean Connery and Christopher Lambert selected Dunvegan Castle as part of the film's backdrop.

While living in London a few years ago, I initiated negotiations for the purchase of Dunvegan Castle, but the castle's lead roof has rendered the site an environmental hazard. Replacement and cleanup was estimated at ten million dollars, so I decided not to pursue the purchase.

On an interesting side note, I may have a distant relationship with businessman, entrepreneur and 2016 presidential candidate Donald Trump, who also hails from the MacLeod family. His mother, Mary Anne MacLeod was born and raised on the Hebridean Island of Lewis, off the west coast of Scotland. She emigrated to the US where she married Donald's father, Frederick Trump.[3]

---

2 http://www.dunvegancastle.com/content/default.asp?page=s10
3 https://www.scottishroots.com/people/donald.php

Mom's father and my grandfather, Isaac Polin, was Ukrainian. His parents immigrated to Philadelphia in the early 1890's. They were jewelers specializing in diamonds. When they fled from their homeland they sewed diamonds into the hems of their garments to help secure their future here.

My grandfather Isaac was a genius at logistics and procurement and certified as an efficiency expert. During the Spanish-American War he worked for a general as his right-hand man. He was taken prisoner and soon won the trust and admiration of everyone in prison—even his captors—because he could find and obtain anything. Due to his connections and vital importance to the war effort, the US negotiated his early release in a prisoner exchange.

My mom's father's brother, her uncle Max Polin, was president and founder of the Cathay Oil and Gas Company in China, one of the largest oil and gas reserves in world. He also represented aircraft manufacturers and sold planes to the Chinese military. In the 1930s, he served as the American Director of the China National Aviation Corporation. Max was also a friend of Charles Lindberg. Max Polin is mentioned in the book *Wings for an Embattled China* by W. Langhorne Bond.

Toward the beginning of World War II, Max was interned by the Japanese in China, but was soon repatriated to the United States. Max's nephew Packard (my uncle) writes, "As a business man, it is said that he traveled more air miles in Asia than any other passenger of his era. Because of his extensive knowledge of China and the Chinese, he was recruited as a CIA Officer and returned to Asia for the balance of the war."[4] After the war, the Chinese government confiscated Cathay Oil

---

4   http://www.cnac.org/polin01.htm

and Gas Company's holdings. Had Uncle Max been able to retain Cathay Oil and Gas Company, he would have been a billionaire the likes of John D. Rockefeller with hundreds of billions to his net worth.

In spite of losing Cathay Oil and Gas Company, Max was very wealthy and when he died in 1958 he left a small fortune to my grandparents. In those days, the daughters in the family weren't considered heirs, so most of the money eventually passed along to her brothers Patrick and Marvin. Uncle Patrick went into the Navy and became the highest ranking officer. Mom also had a sister Nann who did very well in real estate in California.

My mother's father was well liked and highly esteemed by all who knew him. He owned a dry-cleaning business in Hayward, CA, and had owned other businesses in the past. But Mom's parents' true passion was Hollywood. Her dad and mom acted as extras in 99 films in Hollywood, including *Gone with the Wind*. They reminisced that at the end of a day shooting that epic film, Clark Gable would head for his room and call out to the crew, "I'll take a broad and a bottle!" My grandparents always had money, drove new cars, wore fancy clothes, and looked and carried themselves like movie stars.

Mom was christened Billie Marjorie Polin. But she never liked the name "Billie," so after she married, she officially changed her name to "Mia." Mom was an accomplished dancer and performed in the San Francisco Opera and Ballet Company beginning as a child for eleven years. Reflecting back on my childhood, I can still see Mom dancing around the house.

My mother worked in her father's dry-cleaning business as a cashier. One of their regular customers was the man who became my father. One day when he came in to collect his dry cleaning, my mom was wearing a bandana over her head and he reached over the counter and playfully removed it. She was upset, not wanting him to see her hair in disarray.

So the next day he came in and placated her with a stuffed animal—a chicken! This unlikely gift prompted him to nickname her his "Little Chicken." There was about a ten-year age difference between the two. They married just a few months after meeting each other and the following year, my mom became pregnant with me.

My dad, George Austin Plein, Sr., came from a German background and was born in St. Louis, Missouri on April 23, 1924. He was extremely handsome. He had a sister whom I never knew. He served in the Coast Guard before he met Mom. In the Coast Guard, he earned the nickname "Chippy" due to the "chip" he wore on his shoulder. Dad wouldn't take any guff from anyone and was prone to fight. He also taught me how to fight when I was very young.

Dad was quite a talker and had always wanted to be an attorney, but he plied his loquacious gift selling cars instead. He became the number one salesman at a major dealership in San Francisco. His job afforded him the opportunity to drive new cars, and like my grandparents, dad and mom behaved and looked like movie stars.

Dad loved Mom, was very kind to us children, and we had a happy family. After I was born, my parents had my brother Bart and my sister Mia, named after Mom. We lived in Oakland, CA.

When I was in kindergarten, I came home from school one day and saw an ambulance, fire trucks, and police cars outside our home and my young mind thought, "Neat-O, look at all the lights!" All the police officers and firefighters were inside. Then I saw a gurney coming out of the house. I remember being frustrated because I wanted to go into the house to retrieve a left-over Tootsie Pop I had left inside, but no one would let me go in. I was told to stay outside and play with Bart and Mia.

Mom went off with the ambulance and didn't come home for quite a while. When she did come home that evening, she told us that our dad had passed away. I was heart-broken. I couldn't believe my dad was dead and I went into hysterics. I was a lost little boy for a couple years. Dad died on August 15, 1962.[5]

The combination of my dad dying and all the subsequent moves we had to make resulted in my having to repeat second grade. I still look back on my dad's death as one of the most devastating times in my life. He was only 38 and I was 5.

Not until after my dad died, did I learn that he was an alcoholic and that his heavy drinking had brought on his heart attack. Both he and my mom had protected us from that dark side of his life. My mom's parents had known about his drinking and had never thought much of him because of it. Dad's addiction to alcohol and his premature death brought on by alcoholism is the reason that I never became a drinker or drug user. I made a conscious decision at five years of age never to drink or do drugs.

With my dad gone, Mom went to work at a drive-in. We couldn't keep the house, but my grandparents felt sorry for us

---

5 Family Search, California, Death Index, 1940-1997, https://familysearch.
org/search/collection/results?count=20&query=%2Bgivenname%3AGeorge%20
%2Bsurname%3APlein%20%2Bdeath_place%3A%22Oakland%2C%20CA%22~%20
%2Bdeath_year%3A1960-1963~&collection_id=2015582

and bought us a brand new mobile home, paying cash for it. I remember everything in it being shiny, clean and new.

When my dad died, I was the oldest male in the household and felt responsible for the family. I wanted to do the right thing for the family. From that young age I nurtured a strong sense of financial responsibility for my family and still hold to that today. So it should come as no surprise that it was around this time when I started my first business—a Kool-Aid stand.

Sometime later, we lived right down on the wharf in Benicia, California. One day I was outside and noticed that a large barge had broken loose from its moorings and was floating out into the San Francisco Bay. I felt duty-bound to find its owner. So I ran up and down the street to all the local businesses asking who owned the barge so I could let them know it was floating away.

Finally, I found the owner and he was extremely grateful that I had alerted him about his barge. He wanted to give me money as a reward for telling him, but I refused it. Instead I said, "No, I don't want any money, but I sure could use some help getting into business." So I asked him if he would buy me a shoe-shine kit. He bought me a nice kit and I peddled my shoe-shining services up and down the streets of Benicia. I worked hard and did alright with that business. Also around this time, I began buying newspapers on credit and selling them on the street.

Already at this young age, the entrepreneurial spark was beginning to fan into flame.

Chapter Two
— A BUTT-WHOOPIN' —

Whoever spares the rod hates his son, but he who loves him is diligent to discipline him.

Proverbs 13:24

A YEAR AFTER my dad died, my mom met and married Jerome (Jerry) LaMoureaux. He brought two children with him into our family: Susan and Jerry Jr. and then he and my mom had Max. In birth order, we six children are George, Bart, Susan, Jerry Jr., Mia, and Max. My step-dad also had two other children Nick and Janene, but they were older and had already left the nest.

My new dad, Jerry, grew up on a farm in Britton, SD. He was one tough character. He was a big guy and incredibly strong. He and his brothers were so strong, they threw hay bales around on the farm like match boxes. One time before he met my mom, five guys attacked Jerry's brother in a bar in South Dakota and were trying to kill him. In self-defense, Jerry sent a couple of them to their graves and saved his brother's life.

Jerry was a truck driver, mechanic, heavy equipment operator, and carpenter. Even when I was very young, I helped him work on cars—I'd hold the light, get the wrenches, or do whatever he needed. One time he was pulling the engine out of a truck and the engine hoist broke, so he wrapped chains around the engine, stood with one foot on each fender, and lifted the whole engine, heads, carburetor and all out of the truck and swung it out onto the ground! He was as strong as an ox.

My new dad was also a very strict disciplinarian. Coming from a family of 13 children, he didn't pull any punches when it came to discipline! He spanked my butt on a regular basis. When he came home from work I'd run out the back door, knowing that a butt-whoopin' was in store for me. He was real tough on us!

Early on he used a cutting board to spank us until he broke it on me. Then later he'd just bark, "Go get me a stick!" One time I came back with a long, skinny stick about the size of rail on a baby crib. That was a huge mistake, because a switch that size and length really hurt! I didn't make that error twice. I conditioned myself not to cry out or make any noise during a spanking, because that wasn't very manly and I feared I'd get worse.

The saying goes, "Spare the rod and spoil the child." And God knows I often deserved a spanking! But Dad never spanked me in a fit of rage, nor did I ever feel abused. He didn't smoke, drink, or do drugs and he never laid a hand on my mom, but he was tough as nails. I feared him back then. I didn't feel loved by him until later, but I respected him and I soon realized that he was training and preparing me for difficult challenges in life.

My dad always worked hard, but with six kids we never had much money. Even though my mom's parents had been fairly wealthy, as I mentioned previously, their wealth did not filter down to us.

Early on, we lived in a trailer court in Vallejo, CA, and later in the housing projects in and around Oakland (known as the

"Open-Air Prison" today). This region was a hotbed for racial riots in the 60s and murders occurred in the projects on a regular basis. As you read on about my later life, please keep in mind that I was not born with a "silver spoon in my mouth." I grew up in poverty.

During this season of life, I managed to find some wholesomeness by joining the Cub Scouts and later graduated to the Boy Scouts. My mom also provided the more protective, soft, and spiritual side of my upbringing. I had been baptized Catholic and she took us to Mass every Sunday. She was the spiritual leader in our family.

In spite of our humble lifestyle, Mom always made cookies and cakes and kept a clean and beautiful home. I have wonderful memories of her exuberance, dancing and singing, filling our home with joy and grace.

> Growing up with George was kind of like growing up with a master leader. He was always the leader. He always had something going. Everything he did was 100%, whether it was Karate or whatever. That's George. He is a worker!
>
> He's a good brother. He has always tried to inspire me to be a comedian. He focuses his attention on your strengths and tries to bring out the best in people.
>
> He doesn't judge, but finds the best in others.
>
> I'm glad he's my brother and I'd do anything for him.
>
> – Jerry LaMoureaux, Jr.

At dinner time, with all eight of us around the table, the protocol was *no talking*, just down-to-business eating. Mealtime was a serious matter. But one time, Bart, who was ever the comedian, stepped up behind Dad and began making goofy faces that only the rest of us could see. I had just shoveled a healthy portion of food into my mouth and started snickering. When I couldn't contain myself any longer, my mouth exploded in laughter spraying food all over the table.

School, on the other hand, was no laughing matter. I went to elementary school at Flosden Elementary, an all-Black school. We lived about three miles from the school and there was no school bus service. So I walked to and from school every day. The route led me through a ghetto, over a highway, and past a cattle yard and slaughter house. I had to pass through this gauntlet twice a day.

I was the *only* white kid at Flosden. *Every* day I fought out of necessity. I was the *honky*, the *white patty*, the *cracker*. That's what they called me. Naturally, I was the kind of kid who couldn't keep his mouth shut, so I hurled racial insults back at them in retaliation. This of course only intensified their attacks on me.

Most everyone hated me, because I was white. Even so, I held no animosity or hatred towards them. However, my best friend, Duane, also a Black kid, had some empathy for me. The two of us were the toughest kids in the school. I didn't have a choice whether or not to fight. By comparison with the other boys, I was a little guy. Everybody was always bigger than I was, but they weren't as tough as I was due to my father's disciplinary actions. I feared my dad more than I feared the other kids. I simply stood my ground and knew that, as Mark Twain said, "It's not the size of the dog in the fight. It's the size of the fight in the dog."

We fought in the classroom, on the playground, and after school. When we fought or messed up in class, the teachers would spank us publicly with a "board of education." This was a paddle about two-and-a-half feet long with holes in it to increase the speed with which it could be wielded and magnify the pain. Even so, the pain inflicted by the "board of education" paled in comparison with the lickings my dad gave me. These school spankings were laughable.

I had a few friends there, but I fought *every* day. To the extent possible, I was careful not to get myself hurt too badly. Also, as one of the fastest kids in the school, I could get away when I needed to.

One time by the end of the day I had had enough of their name-calling and jabs. I kicked a desk across the room into another kid and called the whole class out. So when the bell rang, we all went outside to have it out. I knew I couldn't fight them all at once, so I ran to thin out the crowd then I turned and fought. Then I ran again and fought again. I continued this pattern of running and fighting the whole three miles back to our trailer court.

By the time I got home, I *still* had some guys chasing me, but now I was fighting on my own turf, though greatly outnumbered. Here's where I had to improvise. I carried a stout, black metal, engineer's lunch box. This lunch box was military grade. I had pushed little wooden pegs into the clasps of my lunch box to keep them from opening when I pounded some guy over the head with it. With the help of my lunch box I started beating the hell out of them.

The reader might wonder what lofty vision motivated me to continue going to school under these awful circumstances. I'll tell you what made me keep going, it was the vision of getting a butt-whoopin' from my dad! He was one tough man. I wouldn't think of defying him and would rather face the fights with my classmates than face him.

Looking back on those years, it was my dad's discipline that made me tough. I figured if I could endure what he dished out, I needn't be afraid of anybody else. Oh, he didn't mind if I fought, but I'd get a licking if I *lost* a fight. His hard word to me, "If you don't win the fight, don't come home," made me stronger. He taught me to fight in order to win. I wasn't big like he was, in fact I was small, but I was strong, wiry, and fast.

I know this behavior probably freaks out some of my readers, but as a little white kid in an all-Black school in the 60s it was a matter of survival. My young life was tough.

During those years, my dad's jobs moved us all over. Just months after the 1964 earthquake in Alaska, Dad moved us to Anchorage where he had worked previously. The first house we moved into was split down the center from the earthquake. You could actually see the sky when standing in the house until Dad repaired it. Our second house was so small that all six of us kids slept not only in the same bedroom, but also in the same bed.

I went to Mountain View Elementary school in Anchorage—not as bad as Flosden Elementary in Vallejo, but still a rough school. We were poor and always seemed to end up in the rougher, rundown parts of town.

Mom and Dad didn't always get along in those years. A couple of times, Mom got fed up with Dad and took off with us kids. But Mom and Dad always got back together and worked things out.

As I mentioned, Mom was the spiritually sensitive adult in the household. But during these years she went through a spiritual identity crisis. She began investigating and trying out many religions: Christian Science, Unity Church, Yoga,

Buddhism and other Eastern religions. She rejected Islam out-of-hand. In the end, she concluded that none of these religions genuinely answered the questions of life. In her view, none of them even came close. She referred to them as "a bunch of hocus pocus religious bull." Only Christ and Christianity provided the answers and peace for which she so fervently sought. So she returned to Christianity.

From 1964 to 1972, we moved back and forth between California and Alaska probably ten times. In California we lived in Vallejo, Oakland, Benicia, Fresno, Stockton, Merced, San Jose, and Sacramento. On one of our trips up the Alcan (Alaska-Canada Highway), my dad, mom, all six kids and a hitch-hiker were in our station wagon. Dad and the hitch-hiker took turns driving while the rest of us slept. Back then the Alcan was little more than dirt ruts.

We had driven into a snow storm and the hitch-hiker lost control of the car and put us in the ditch. The accident broke the rear axle. With help we got the car into Watson Lake, British Columbia and stayed there for several days while Dad repaired the car.

One day during our stay I went to the grocery store for some candy. Just as I arrived at the door, the store manager kicked a Native woman out of the store and onto the ground. His treatment of her angered me so much that I jumped on him and beat him up—even though I was just a kid of perhaps twelve. As a kid, fighting was the way I had learned to solve disputes.

My dad used to remark, "People say that George will fight at the drop of a hat. But George doesn't wait for the hat to drop!" It's true, I had quite a temper back then and it didn't take much to ignite it.

One summer, my dad took me to Britton, South Dakota, to work with him. We worked on the family farm picking rocks, cleaning the pigpen and doing other farm work. At other times in Alaska, Dad bought buildings that were slotted for demolition. We would dismantle these old buildings, careful to reclaim all the wood. We pulled the nails out of the lumber, straightened the nails for reuse and sorted and stacked the lumber. Then we took the salvaged lumber to another location and built a new house with it. Dad did whatever it took to make a living and he worked very hard. And when I worked with him and for him, I had to work my butt off!

Dad's example rubbed off on me and I grew up loving hard work. On one of our stints in Alaska, I worked a morning paper route as a kid. I had to get up at 4:30 am to start delivering papers at 5 am, no matter how cold or snowy it was. Once, in the dead of winter, I was out at 5 am delivering papers. The morning was cold and pitch-black. Suddenly, an asteroid or meteor flew overhead lighting up the sky as though it were daylight! Then it was gone. I never found out exactly what it was.

In my teen years, it was as though a switch flipped in my relationship with my dad. Suddenly, he became very supportive and proud of me and my accomplishments. My memory of him since then is of a warm, compassionate, caring, good man trying to do the right thing for his big family. He has always been there for me. He co-signed for me when I bought my first new car, a '77 Monte Carlo and he was proud of me when I paid it off early.

Dad's example of hard work and discipline really began to pay off in my life when I reached my mid and late teens. In fact, I must credit him greatly with much of what I began to accomplish in those years.

Chapter Three
— TOUGHER THAN BOARDS AND BRICKS —

If you always put limits on everything you do, physical or anything else. It will spread into your work and into your life. There are no limits. There are only plateaus, and you must not stay there, you must go beyond them.
Bruce Lee

IN MY PRE-TEEN years, Dad taunted me, "You can't do it!" To which I would stubbornly reply, "Oh yes I can!" Whether Dad knew that's what I needed to take initiative and build perseverance, or whether things just worked out that way, his methods served me well.

Consequently, when I was 14 or 15 years old I decided to improve myself and expand my capabilities. In order to facilitate my self-improvement, I attended the Dale Carnegie course and completed Paul Meyer's Success Motivation Institute (SMI) through books and cassettes.

Dale Carnegie's course sets out "to inspire and motivate people to a higher level of performance." And Paul Meyer's famous quote became the compass for my life. "Whatever you vividly imagine, ardently desire, sincerely believe, and enthusiastically act upon... must inevitably come to pass!" Both of these experiences, coupled with my dad's example and training set me on a path to make something of my life and change the world.

In my early pursuit of business skills, I also became acquainted with the *Don't Quit Poem*. The message of this poem has also had a profound impact on my life. Over the years, I've woven its truths into the fabric of my life. This poem is part of who I am today.

Don't Quit Poem

When things go wrong, as they sometimes will,
When the road you're trudging seems all uphill
When the funds are low and the debts are high,
And you want to smile, but you have to sigh,
When care is pressing you down a bit,
Rest, if you must, but don't you quit.

Life is queer with its twists and turns,
As every one of us sometimes learns,
And many a failure turns about,
When he might have won had he stuck it out;
Don't give up though the pace seems slow–
You may succeed with another blow.

Often the goal is nearer than,
It seems to a faint and faltering man,
Often the struggler has given up,
When he might have captured the victor's cup,
And he learned too late when the night slipped down,
How close he was to the golden crown.

Success is failure turned inside out–
The silver tint of the clouds of doubt,

*And you never can tell how close you are,
It may be near when it seems so far,
So stick to the fight when you're hardest hit–
It's when things seem worst that you must not quit.*

— *Author Unknown*

During my teen years, we continued to move back and forth between Alaska and California. I attended parts of elementary, junior and senior high school in both states. High school in Vallejo, CA was just as tough as Flosden Elementary had been. In every school I attended, I had to prove myself in order to earn respect and that respect came only through fighting. Consequently, my frequent change in schools led to an increase in fighting.

Because we always wound up living in poor neighborhoods, I attended rough, inner city schools that lived up to their reputations. In California, we experienced non-stop gang-fighting, drug-busts, racial wars, and police interventions. This was all part of my education and formation of who I became.

At 15, I was drawn to martial arts and Karate in particular for several reasons. First of all, I continued getting into fights at school, but I was 5'9" and light weight (even though I was two inches taller than Bruce Lee who stood at 5'7"). I had to be faster, tougher and win in my fights, or it would be a bad day for me. I thought that martial arts would help me compensate for my stature.

Also, I've always been inspired and captivated by television and movies. About this time, the television shows *The Green Hornet* starring Bruce Lee as Kato, and *Kung Fu* with David Carradine aired. Then in 1971 Bruce Lee's first full length movie *The Big Boss* was released and two years later, Lee's movie *Enter the Dragon* hit the big screen. Watching these shows and others further fueled my desire to pursue Karate.

Finally, I simply had the drive to continue to improve myself and believed that the discipline of Karate would help me do that. So, beginning in California, I threw myself into Karate with fervor. My small, but strong and lean frame was well suited to the sport and my mind and body had been well prepared for the discipline. Because my dad had driven me so hard when I worked with him, I found Karate, and nearly everything else I tried, easy by comparison.

After establishing a good foundation in Karate in California, we moved back to Anchorage and I continued my training at Tanaka's Martial Arts Academy. I trained under Sensei Tanaka for about two years. Tanaka ran a great school, but after participating in a martial arts competition where I saw Robert Alejandre compete, I was so impressed with his skills that I asked if he would train me.

Robert Alejandre was (and still is) an incredibly gifted martial artist who had placed overall in the International Karate Championships in Amsterdam. None other than Bruce Lee had presented Alejandre with his trophy. Alejandre had studied Go Ju Ryu Karate under Goshi Yamaguchi, the son of the famed Sensei Gogen Yamaguchi ("The Cat").

Robert had just finished his tour of duty with the military when I met him. The US Army had capitalized on his expertise as a martial arts instructor, employing Robert to train military personal in the martial arts under the command of Brigadier General Stephen A. Cheney.

Alejandre is a five-foot-seven, incredibly dangerous man. In the '70s, while the Alaska pipeline was being built, Alaska was a very rough place. There was also a strong military presence at Elmendorf Air Force base and Fort Richardson Army base just north of Anchorage. During this period, a steady flow of powerful farm boys from the pipeline and Special Forces military personnel would come into the Dojo to challenge Alejandre.

Alejandre had quite a reputation for his skills, but he did not invite this kind of attention. But as men everywhere and for all time have sought to prove their masculinity by besting others, in the same way these tough brutes came to the Dojo thinking to walk away having beaten Alejandre.

Typically, these glory seekers were much bigger than Alejandre. They rippled with muscles and a lofty opinion of themselves. Sensei would coolly receive them and let them make their first move. Often he would play with them like a cat with a mouse, and then he'd pull some amazing move on them and before they knew what hit them they were laying on the ground moaning.

> I had just gotten out of the service when I met George. He and his brother Bart came to be instructed by me in martial arts. I initially taught George in his basement, then in my school when it opened in 1980.
>
> George had a tremendous amount of inner energy as a martial artist—more than anyone I've ever met. He was among the top four or five students I've ever taught. His black belt was not handed to him. He worked very hard to earn it.
>
> George and I became like brothers and are close to this day. We used to run and work out together. Then we'd go out on the town. George is a great dancer and is a lot of fun to be with. He always does everything with zest!
>
> – Robert Alejandre

Alejandre would go over and coddle them, "Oh, I'm so sorry! Are you okay?" It was all I could do to stifle my laughter! I knew exactly what Alejandre was doing with these guys to humble them. I never saw anyone beat him.

Sensei Alejandre ran a very non-traditional, unorthodox Karate school. Alejandre's Martial Arts Academy was in the heart of the Anchorage ghetto. We did not train on mats, but on the bare concrete floor and without pads. It had been much the same at Tanaka's.

While Alejandre ran us through our Katas, kicks and exercises, he walked among us and whacked us with a bamboo stick. The purpose of this exercise was to train us to maintain our form regardless of the distraction. The butt thrashings my dad had given me had conditioned me well for this punishment. Alejandre often demanded 500-600 kicks and several hundred punch combinations in a session to ensure we didn't break form.

It took me six years to earn my black belt in Karate from Alejandre. Alejandre was also unconventional in his approach to how he awarded that distinction. He required me to fight everyone who came into the Dojo! My matches included fighting those tough military Special Forces soldiers and challengers from other Dojos. Over the years I've competed in over 700 matches. These were full-on, free-style matches, but they were always a scaled back version from what Alejandre taught us, for our intent was not to kill or maim anyone.

Alejandre's philosophy was that as a black belt, you didn't lose a match. You had to win. You had to be "bad"—not deadly, but incapacitating. We always fought clean without intent to cause injury, but sometimes it couldn't be helped. With black belt status also came responsibility and liability. Mastering martial arts is not all physical, but also highly psychological involving a lot of reading, study and mental discipline.

After receiving my black belt, I began training in Mixed Martial Arts. I fought other martial arts schools including Yang's Tae Kwan Do School. Grand Master Woo Yup Yang was an incredibly tough martial artist and world-renowned for his skills.

One time Alejandre and Yang agreed to an inter-dojo match in which our best martial artists competed against their best. I was pitted against their best man and won the fight with ease. But Yang did not work his students as hard as Alejandre did and we annihilated them! Though not intentionally, we sent several of their students to the hospital, one or two of which I was personally responsible for.

Alejandre also invited world renowned martial arts experts to cross-train us in their areas of specialty. Some of the masters I learned from included, Benny "The Jet" Urquidez, a kick-boxer and full-contact fighter who held six World Titles in five different weight divisions. Another of our guest instructors was Bill "Superfoot" Wallace, the Professional Karate Association (PKA) Middleweight Champion kick-boxer for nearly six years. "Superfoot" Wallace owned the distinction of being the world's fastest kicker in the martial arts. I also studied under Eric Lee, Grandmaster of martial arts, dubbed the "King of Kata".

Sometime later I was inducted into the PKA. I used to break up to five bricks and boards set on fire. I did this by executing flying side-kicks, Karate chops, punches, head-and-elbow breaks, and with open-handed palm punches. Here's what a typical martial arts demonstration might look like at a mall, school, college, etc. First, I came on stage and broke boards with my head while one of the other black belts held the boards directly in front of him.

Then, I had positioned three martial artists to my left and three to my right holding bricks. The bricks were two-and-a-half inches by nine inches by twenty four inches. Two people held the bricks with the third stabilizing those two. Without support of this third person, my kicks to the bricks would have knocked the other two individuals down. I executed each break left and right with flying sidekicks.

Next, I approached three stacks of wood supported by bricks, with five boards in each stack. These I broke with a Karate chop, punch and elbow. In the grand finale I broke five bricks set on fire using Sterno and shredded newspaper creating an inferno. When I hit those blazing bricks with an open hand flames jumped nine to twelve feet in the air!

Kung Fu Magazine and ABC, CBS and NBC television networks featured me performing these board and brick breaks. While impressive for an audience, the purpose behind these stunts was to ensure absolute follow-through and commitment. But as Bruce Lee once observed, "Boards and bricks don't hit back."

As my proficiency and experience in Karate and Mixed Martial Arts increased, I began teaching Karate at the University of Alaska in Anchorage (UAA). With Alejandre, I co-authored the Women's Self-Defense Handbook. At this time, US Olympic officials approached Alejandre and he invited me to train for the US Olympic Black Belt team. I seriously considered this opportunity, but at the time, it didn't make sense to focus on a non-income producing pursuit with the Olympics. Instead, my chief focus turned toward business where by now I was making millions of dollars per year.

Chapter Four
— A PIONEER, NOT A FARMER —

Whether you believe you can do a thing or not, you are right.
Henry Ford

WHILE I PURSUED my business career, I continued my martial arts training and workouts for many years to come. Martial arts was not a fad recreation of mine that eventually receded into my past. Karate became integral to my life and who I am. And little did I know then how useful those martial arts skills would prove in the future.

No matter what I was doing, I always looked toward the next challenge and I was never content to focus on only one thing. During these many years that I actively pursued martial arts, I also began to develop as an entrepreneur and businessman. At 15 I had taken a job with a Texaco station and carwash in Alaska and worked long hours. In peak season I was working 108 hours a week at $2.10 an hour, straight time. By those days' standards I was making a lot of money. And this work was easy compared to working for my dad!

While employed at the Texaco station I paid cash for my first car, a '64 two-door, lemon yellow Chevy Malibu. The ground thundered under me as I cruised along in this street-legal competition race car. My Malibu turned a 10-second quarter mile with its 396 cubic inch engine fitted with 427 heads, Mickey Thompson headers, Holley carburetor, Isky cam, four-five-six gears with a Hurst shifter, a 411 rear end, Cragar wheels, and wrinkle-wall slicks.

Equipped as it was, my Malibu could pop a wheelie! The car had been owned and professionally raced by Mr. Kolslosky, owner of Kolslosky's Men's Store. He had raced the car at the Polar Drag Way in Palmer, AK. This Malibu was an impeccable showpiece. Unfortunately, I may have tested its limits one too many times and the engine blew so I sold it.

During those days, automobiles were my first love—then girls. Sometime later those two passions flipped in priority. More recently however, I've thought about going back to cars. Even so, I keep my eyes open for Mrs. Right.

My family and I went back to Vallejo, California when I was 16. There, I bought a red '69 Pontiac GTO—a gorgeous car with black interior. In Vallejo, I had a steady girlfriend, Joy, who was two years older than I. I really thought I was in love with her and that we would eventually marry. Then one day Joy jilted me. She said she was leaving me for a doctor who had lots of money—something I did not have at the time. Even though I had a job, I couldn't compete with a doctor.

I was shocked and incredulous! "What does money have to do with our relationship?" I blurted out to her. But she left me. I was quite broken-hearted, because she meant a lot to me. But from that day forward, I promised myself I'd never be without a substantial amount of money again. Sometime later, she wanted to come back to me, but by then I didn't want any part of her fickle affections.

Later that year, we moved back to Alaska. I traded my GTO for my brother's Chevy Van with mag wheels and drove it up the Alcan at 16. Moving all the time and constantly changing schools was tough. I fought all the time after school whether I wanted to or not.

During these years in Anchorage I worked at McMann's Furniture and achieved the second-highest sales in the nation. I also worked at Alaska Maintenance Contractors—a janitorial service, and at Holiday Motors selling cars for a short time. Another place I worked was Don Hoffman's Construction company where I learned to build houses from the ground up. His son, Mike, was one of my best friends in high school. Don drilled into me the mantra, "Always think ahead." I worked my butt off at all of these places and did very well, making lots of money.

One day I was driving by a car dealership and saw an immaculate black '69 Cadillac Eldorado with red leather interior. The Cadillac caught my eye and I decided I had to have it. At the time, the Eldorado was the most expensive car I'd ever purchased. The car had belonged to the owner of the dealership and it became my prized possession. I paid cash for it.

At 17, I served as president of the East Anchorage High School Junior Class. My life was focused on Karate, cars, girls, and work. Then one day I saw an ad in the paper that Alaska Maintenance Contractors (one of the companies I had worked for) was for sale. I made some calls and using my Cadillac as a down payment I purchased that business. In exchange for my Cadillac, I received a Volkswagen van equipped with janitorial equipment and client contracts. My new business venture squeezed my time to the max, so I left high school after my junior year.

I changed the name of the company to Alaska Maintenance and Janitorial (AMJ) and soon found that I had a knack for business. During AMJ's first year of business, dollar volume quadrupled. Circumstances were also on my side. The majority of the workforce had left town to build the pipeline, so I "cleaned up" Anchorage. I bought out Townsend Janitorial and added their contracts to ours. As part of AMJ, I also built Magic Clean Carpet Cleaning. Within three years I had 54 employees and had established contracts with the State buildings, the hospital, grocery stores, schools, banks, hotels, and construction cleanup.

Nearly everyone who worked for me was older than I. At first, they thought I was just a kid. But I earned their respect when they saw my professionalism—the crisp uniforms, established processes, clean equipment, and great service.

> George and I were students at East Anchorage High School when we met. George was a smart dresser. He always awed us with the fact that as a high school student he was already the owner of a janitorial company! He and I have been great pals ever since.
>
> – Patrick McCourt, Entrepreneur and friend

I was making about $150 thousand a year in my business. And in the fall of '76, I bought my first new car, a '77 burgundy Monte Carlo with white leather interior and 10 inch mag wheels and a car phone!

AMJ was my first real business and I discovered an important distinction about myself. Perhaps I can best describe it in the form of a story. One man comes over a hill and beholds a pristine valley. In his mind's eye he envisions its great potential and all that might be developed in this valley over time. Another man comes over the same hill after the valley has been developed and says, "This is where I can settle and establish my family for generations to come." The first man is a *pioneer*, the second a *farmer*.

What I learned about myself is that I'm the *pioneer*. I envision what could be. I love designing, launching and building a business. I'm not satisfied with the status quo. I'm always looking to improve on what exists, or invent what no one else has thought of. Also, I love learning new things. If I don't know how to do something, I throw myself into researching it until I become an expert at it.

After three years of building up AMJ I was ready for a new challenge. I set my sights on purchasing a restaurant franchise and was able to obtain the A&W Root Beer Restaurant franchise from Leon Brown who owned this state-wide franchise at the time. I then selected some prime property on which to build.

Next, I approached multi-millionaire real estate developer Kris Gratrix and proposed that we go into partnership together to build what would become the world's largest A&W Root Beer Restaurant franchise. Kris agreed to the idea and pulled in his partner, a banker named Ray Hartlieb. Initially, I continued to manage AMJ on the side. The plan was for me to build and operate the A&W franchise, while Kris and Ray financed the $1.8 million loan for this new restaurant.

In 1976, on my own initiative and at my own expense, I flew to Pearl City, Waikiki, Hawaii in order to visit and observe A&W's highest volume restaurant (at the time). Meanwhile, in Anchorage, we began construction on our brand new A&W Root Beer Restaurant. Our new restaurant had 12 cash register stations at the order counter. I sought to innovate at every turn.

Consequently, I helped invent the preset-button system on cash registers to speed up orders and the batch order system that Burger King later adopted.

Our restaurant soon became the largest, highest volume A&W Root Beer Restaurant in the world. At one point we had a group of McDonald's executives sitting in our restaurant with stop watches trying to figure out how we were filling orders so quickly. From time to time, former Governor Eagan would come to the restaurant and eat lunch with me. He was the first governor of Alaska, but served a second term in the '70s. We were good friends. I have also known a lot of the other governors of Alaska over the years.

In 1978, in view of our huge success, my business partner became greedy. He had insisted on 51 percent ownership at the outset. With his majority ownership, he decided he wanted to charge me $15,000 per month rent, twice what the property was worth according to the MAI real estate appraisal performed by Erickson and Associates. (They assessed its lease value at $7500 per month.) So we broke up our partnership and he bought me out. As it turned out, he couldn't run the restaurant and ended up selling the location to Burger King where it still operates today. But this experience whetted my appetite for the restaurant business.

Chapter Five
— "CHICKEN GEORGE" —

I have always been driven to buck the system, to innovate, to take things beyond where they've been.

Sam Walton

WHILE OPERATING THE A&W Root Beer Restaurant, I had sold AMJ for a substantial profit to Ed Sanders, who turned it into Nova Property Management—still a profitable business today. In this transaction I received two pieces of property and a large amount of cash. Those assets, coupled with the sale of my portion of the A&W restaurant, enabled me to purchase two Church's Fried Chicken franchises.

With the acquisition of the Church's Fried Chicken restaurants, I flew to San Antonio, Texas to attend the Church's Franchise Training School. When I returned to Alaska, I remodeled both restaurants. These restaurants did very well, earning me the Vince Lombardi award for the highest sales volume increase in the nation.

My first memory working for George was at Alaska Maintenance and Janitorial. I was only 11 years old! I was half a man, so I received half a wage. I learned to sweep, mop and clean carpets.

I learned a lot from my older brother and followed him around to his other businesses. At 15 I managed all his Church's Fried Chicken restaurants and had my own apartment.

– Max LaMoureaux

I had a knack for starting businesses and my brother Max was skilled in managing them. Although Max jokes about receiving half a wage at 11 years old at AMJ, he soon became my highest paid employee managing my restaurants. My brothers Bart and Jerry worked in my restaurant business as well and my dad made our daily deposits.

Our first two restaurants were doing so well that I decided to launch two more Church's Fried Chicken restaurants. To facilitate this, I bought out two Pioneer Chicken restaurants, one in Fairbanks and the other in Anchorage. I closed the one in Fairbanks and brought the equipment to Anchorage. This gave me four Church's Fried Chicken restaurants in Anchorage.

Next to one of my Church's Fried Chicken restaurants, I owned some property and decided to build a Taco John's restaurant there. Then, I opened a sixth restaurant called Mama Mia's Deli, named after my mom.

In addition to these six restaurants, I also launched a limousine service called Rich Limousine. Rich Limousine was truly a family business that included Max, Mia, Bart, Jerry and me. The Alaska Journal of Commerce published an article about Rich Limousine, touting ours as the longest limo in Alaska![6]

In a crazy set of circumstances, I found out that country singer Kenny Rogers tried to finagle the brand new white Lincoln that we had ordered. In the end we got the car and he had to wait for the next one to come off the line. I also owned a snow-plowing business at this time. Now in my mid-twenties, I owned businesses worth millions of dollars.

---

6   Nancy Cain Schmitt, "LaMoureaux Offers Longest Limo in State," Alaska Journal of Commerce, Vol. 8, Nr. 43, October 22, 1984, cover and page 10.

One of the Church's Fried Chicken locations was not doing well due to local demographics and its unique impact on selling chicken. Selling chicken requires a higher population density than selling burgers. So we converted that Church's Fried Chicken restaurant into Max's Beefy Burgers. Max's Beefy Burgers took off and became profitable and is still in business today.

During the restaurant years a mutual friend of ours introduced me to Carol. She was a stunningly beautiful woman; so much so that she became a model for the John Casablanca Modeling Agency in New York. Sometime later she took over the Barbizon Modeling Agency for the Western half of the US including Hollywood. Carol posed as a Covergirl. She looked like a young Jane Fonda or Shakira. Wherever we went, when I entered a room with her every head would turn.

Another of Carol's distinctive characteristics is that she is the daughter of an Air Force General. I eventually married the general's daughter and we were together for seven years, but had no children. Carol fit right in with my family and especially loved and was loved by my mom.

George was my first love. We had lots of fun together. We were very young when we met and very fond of each other. In many ways we grew up during our years together. I was a little fire ball; he was always dreaming of the next thing.

George taught me how to be an entrepreneur. He and his family had a huge impact on me.

George was always figuring out ways to help others. He was very giving and loving—he was inspiring to be around. He is one of a kind—there's nobody quite like George.

– Carol Christian, Alternative Health Care Practitioner and Founder Eternal Health and Wellness

Mom taught Carol to oil paint and belly dance. Together they taught Middle-Eastern dancing at the Teamster's Hall in Anchorage. Mom also served as Carol's mentor introducing her to holistic medicine. For the past 19 years Carol has owned her own business operating health and wellness clinics in Reno, Las Vegas, and now in Brentwood, California.

During those years with Carol we had great fun together. I was making a lot of money and we spent some of it together on exotic vacations. I recall one time we decided to vacation in Hawaii. I've been a swimmer all my life and had earned two scuba diving certifications. One day, Carol and I hired a boat to take us out into deeper waters to go diving. On one dive, I was swimming near the bottom when movement caught my eye. There, not far from me a giant moray eel slithered between two rocks. I judged its length at 10 to 12 feet and its girth about 18 inches! After observing this eel for a while and keeping a wary eye on it, I decided to swim for the surface. My restaurant franchises enabled us to vacation in glamorous places and stockpile fond memories like these.

As owner of three Church's Fried Chicken restaurants, my friends affectionately called me, "Chicken George." This was a play off the character "Chicken George" in the 1977 television series *Roots*.

However, after working those six restaurants over a six-year period, I began getting antsy. Fast food restaurants demand a great deal of hard work and no prestige. Although I was making a lot of money, there was no glamour in owning a chicken restaurant. My restaurants earned me a lucrative income, but they weren't *cool*. Money is overrated. To me money is just a tool. I was beginning to feel too much like a *farmer*. I needed to pioneer something again.

So in 1984 at the age of 28, I sold all six restaurants and paid off all unpaid debts. With my profits I went into the nightclub business. First, I bought a downtown bar called The Monkey Wharf featuring about 40 monkeys in a glass cage behind the bar.

When I bought The Monkey Wharf, it had already been a profitable bar, nightclub and strip joint, but I had no intention of running a strip joint. So shortly after I bought it, I got rid of the monkeys and the strippers. I decided to create an entirely new environment, so I gutted The Monkey Wharf and refurbished it elegantly and named it The Ritz. I continued running the business during construction in order to pay the bills.

The Ritz was designed as an upscale nightclub—a disco-tech—and restaurant done up in a "Roaring '20s" theme. The male personnel wore tuxedos and spats, while the female personnel wore flapper dresses. My office was behind the bar with one-way glass so I could oversee the whole place. This was a multi-million dollar facility and the number one happening place in Anchorage.

Pictures of the movie *The Sting* graced the walls. The Ritz boasted marble floors, cut crystal doors, and granite counter tops. Some of our chairs cost as much as $1300 a piece. The Ritz shone cold-to-the-bone spectacular! At the time it was the most opulent establishment ever built in Alaska. This club was so successful that I decided to purchase a second club, Spenardo da Vinci's in midtown. It was a two-and-a-half million dollar nightclub with a laser light show. This nightclub I transformed into Bogie's, named after Humphrey Bogart.

My nightclubs catered to politicians, big business, movie stars, and Playboy and Penthouse celebrities. I remember meeting NBA Champion Charles Barkley and Teamster leaders at the club. Any big wig flying into Alaska came to my clubs and spent their money. My clubs were *the* place in Alaska to see people and be seen. In Alaska, oil was flowing and so was the money—lots of it. My businesses were thriving beyond belief. On any given night I had 2,000 visitors combined at both my clubs. At closing each night, we carried the cash away in black plastic garbage bags there was so much of it!

With so much money coming in, my nightclubs fueled my passion for exotic automobiles: Mercedes Benz, BMWs, Corvettes, and Porsches. I also owned a five-story house on exclusive Hillside, and never lacked for a girlfriend. I treated the women I dated well and genuinely loved them. I sought after a long-term relationship with a woman.

I also lived in a downtown penthouse. This top-floor condominium was exquisitely designed and appointed. Walking into my penthouse, I could snap my fingers and the lights would come on. Corian® countertops illuminated underneath and solid brass sinks graced the kitchen and bathrooms. My big-screen television appeared out of the ceiling at the flip of a switch.

Impressed as I was with this lifestyle and the prestige it brought me, I have since learned that "Life does not consist in an abundance of possessions."[7] So too, prestige can bloom overnight and wilt just as rapidly.

---

7   Luke 12:15 NIV.

Chapter Six

— WALLOWING WITH THE PIGS —

*There is a way that seems right to a man,
but its end is the way to death.*
Proverbs 14:12 ESV

FOR A FEW years I maintained this very hard-core lifestyle. But let me be clear that this is not a lifestyle I would repeat. The after-club parties, superficial relationships and all these material belongings left me very unfulfilled. My nightclub environments often forced me to interact with the lower elements of society. Getting into the nightclub business was truly one of the biggest mistakes of my life. "If you wallow with the pigs, you start smelling like one."

Back then my approach to women was what I call the "S.I.N. (Safety in Numbers) method." I dated a lot of women, but my intent was always to settle down into a relationship. When I had a steady girlfriend I wouldn't cheat on her. In Alaska in those days, depending on the oil pipeline, commercial fishing and other factors, there were anywhere from 8 to 35 men to one woman, so the women were spoiled in my opinion. As a result, a woman might flit from man to man, because she could.

Being in the bar business, all too often, men would cry on my shoulders whining about the fact that their girl had just left them. My Safety in Numbers strategy was a self-protective response to their stories and my own experience as I reflected back on Joy leaving me for that doctor.

Beyond the baggage that my own lifestyle carried with it, I recognized that sin ran rampant in my clubs. The nightclub atmosphere promoted drugs, alcohol in excess, illicit sex in every form, and all the other vices that accompany these practices. Also, along with the rich and famous, my clubs attracted the more depraved and corrupt elements of society.

With so much money flowing in Alaska, the mafia had moved in and tried to get their claws into me.[8] They wanted to put their vending machines in my clubs and they wanted pieces of my business—both of which I adamantly refused. My closest staff members and I wore bullet-proof vests and toted pistols and shotguns. From time to time, the mafia would send their goons into my clubs and do their drive-bys, but I just kicked them out.

During those turbulent years, two of my cars were firebombed and an explosive device was hurled into one of my clubs, but fortunately failed to detonate. These gestures were meant to instill fear in me to play along with the mafia, but I wasn't about to let anybody intimidate me.

We knew exactly what kind of environment we were creating, that's why we scanned everyone coming into The Ritz with metal detectors. An article in the Anchorage Daily News on October 17, 1985, quotes me saying,

---

[8] By "Mafia" I'm not referring to an extension of the East Coast Mafia. But to a cartel of less than honorable "business men" who wanted to control and profit from others' hard work.

Everyone is welcome. But their weapons are not. My point is to protect my patrons and my employees from anyone who wants to drink and get stupid with a gun. If we have a problem in here, it's not going to be with guns. Most men today are pansies. If they've got a problem, they're afraid to go hand-to-hand. The first thing they do is pull a gun or a knife. But not in here![9]

Later the same article goes on to mention that police were investigating a shooting that occurred in a parking lot next to The Ritz on the previous Saturday. This may be in reference to an incident in which I became intimately involved.

One night I was working in my office at The Ritz doing the books when one of my bouncers, Gene Bollig, the 1982 Mr. Alaska Body Building Champion, came in and informed me that a drunk GI was dancing on the bar naked. Consequently, the bouncers didn't want to touch him. I instructed a couple of my guys to get him down and take him to the restroom along with his clothes.

On the way into the restroom, the GI yelled at a big Black man, "You owe me 50 bucks! You bet me $50 that I wouldn't dance on the bar naked, but I did. So you owe me!" The Black man defiantly spat back, "I'm not payin'!"

I could see that a big fight was about to ensue if I didn't step in, so I told this Black man to come with me into the restroom along with the GI and my bouncers. I told this Black man, "You made a bet so you're going to pay him." When the Black man refused to pay, I had everyone else leave the room.

---

9 Larry Campbell, "Keeping a Closer Watch at City Bars—Club Owners Try to Keep Trouble Out," Anchorage Daily News, October 17, 1985, p. A1.

I stationed my bouncers outside the restroom door and asked them to lock the door. So now it was just the two of us inside.

I tried to reason with him, "Look we're going to have a race riot between you Blacks and the GIs out there and it's going to cost me an infraction on my liquor license. So man-up and pay him the 50 bucks you owe him!"

With that, this guy got violent, looked down on me and said, "Who the f—- are you?!"

I said, "I'm the guy who's going to kick your ass if you don't pay the 50 bucks." And then he swung at me. I sprang into action, applying my martial arts training and in the next few minutes "introduced" his face to the urinal, the wall, and the door. Finally, I marched him out of the restroom, out the back door of the club and into a snow bank. I wasn't trying to critically wound him, but I wanted to send a strong message and give him a chance to leave without causing further problems.

The next thing I knew, the GI had run out the front door of the club and the Black guy had run around to intercept him. The Black man now had a semi-automatic pistol that he apparently had grabbed out of his car. He began shooting and the GI took a bullet to the groin. Then the Black guy jumped in his car pulled around in front of the building and aimed his gun at me.

Insanely, I had no fear and began taunting him to shoot me. I was wearing a bullet-proof vest, but couldn't guarantee he'd hit my torso if he shot. I began working my way toward the car with the intent to disarm him, but he sped off before I could get there.

At this point, I picked up the GI and carried him across the street to the Police Station. We had called the police 20 minutes earlier and they still hadn't responded. I was pretty ticked off at the cops and told them so. They called the paramedics who came and attended to the injured GI. To my knowledge, nothing was done about the man who shot him. That's just how it was in those days.

On another occasion, one of my bouncers ran into the club, found me and blurted out breathlessly that a pimp had one of my female employees in his car with a knife to her throat threatening to kill her. I had a very strong sense of protection for my employees, especially for the women.

I ran outside and saw the pimp in the driver's seat of his car. My employee was in the passenger seat next to him. He had her head pulled back by her hair with a knife at her throat. I bolted toward the car, reached in and grabbed his hand with the knife to disable him. In the same move I yanked him bodily through the car window maintaining a lock on his hand and broke the knife loose from his grip. Then I "put him to sleep" slamming him against a concrete wall.

I was making lots of money and I had the skills to handle these unfortunate incidents, but my clubs were destroying people's lives—including my own.

During this period of time I experienced a troubling incident involving Mike Hoffman, a good friend of mine, and his brothers Ron and Don. The Hoffman brothers owned Custom Coach Auto Body shop. A customer of theirs, Adolf Lerchenstein, became extremely angry over a petty charge and jumped in his truck and began backing up, unaware that he had struck my friend Mike. Mike was now holding onto the rear bumper of the truck to prevent himself from being run over.[10]

Meanwhile, his brother Don, seeing what was happening to his brother, ran to the passenger side of the truck and opened the door jumping in, in an effort to tell Lerchenstein that he was running over his brother. Lerchenstein took his actions as a threat and picked up a gun off the seat and pointed it at Don, but did not fire. Don slid out of the truck.

Seeing what was going on, Ron ran to the driver's side of the vehicle and reached in to grab Lerchenstein. Lerchenstein shot Ron in the chest, fatally wounding him.

I had some business to take care of at Custom Coach and had also come by to see my friend Mike. I arrived on the scene as all this was happening. When Lerchenstein started shooting, I tackled Mike and took him to the ground to avoid being hit, no doubt saving his life.

As you can imagine, this was a very tumultuous time for me on many levels.

---

10   Justia, "Lerchenstein v. State," March 24, 1989, http://law.justia.com/cases/alaska/court-of-appeals/1989/a-2172-0.html. .

Chapter Seven
— FROM HIGH LIFE TO NO LIFE —

Pride goes before destruction, a haughty spirit before a fall.
Proverbs 16:18 NIV

UNFORTUNATELY, I DIDN'T learn until later in life that you don't have to engage in every fight you're invited to, but it was a big part of my upbringing. My dad raised me with the understanding, "If you get in a fight and don't win, don't come home."

In the spring of 1987, the mafia got involved in a plan to build the largest mall in Alaska. I had a meeting with their leaders in my office that brought us to a stand-off. They offered me an incredible amount of money to bring me under their control, but I refused it outright. They also threatened me and my family and I explained to them in no uncertain terms what I would do to each one of them if they ever attempted following through with their threats. If the mafia had attacked my family I would have retaliated to the fullest extent. So in this meeting, I made it clear that I would not tolerate any threats or violence from them. I was fearless back then.

Following that meeting the mafia went about their scheme to harm me in a very devious way. Wielding their influence, they were able to take away all my downtown parking for a distance of six blocks from my club. Their deed cut into the business at The Ritz significantly. Without sufficient parking The Ritz was doomed!

My other club, Bogie's, was still doing well until the city council decided to place a restriction on my liquor license. Such action was unprecedented. Mine would have been the only restricted license in the state of Alaska. I already had two-and-a-half million dollars invested in the club and a half-million in the liquor license and now they wanted to devalue my liquor license. Their stipulation made my license worth hundreds of thousands of dollars less. I felt like I was being backed into a corner financially with my debts looming over me and I became desperate.

So I hired the former Supreme Court Justice of Alaska who was now an influential, hired-gun attorney. He counseled me to act in defiance of the restrictions the city council was trying to put on me. He took them to the mat legally and told them this is what I was going to do and the city council did not follow through on their threats. They knew what they were doing was wrong.

Under these circumstances, I compromised my morals in a direction I had never originally intended. But I reasoned, "Perhaps if I turned both clubs into strip joints they might bring in the cash I needed." So that's what I did. It was really a simple numbers game. Financially, it made sense to turn The Ritz into a strip club, because I could run a strip club with fewer clientele than it took to run a nightclub and restaurant. I knew this to be true, because when I had purchased the Monkey Wharf, it had been a strip joint and I had all their financials. Also, turning Bogie's into a strip club prevented the City Council from restricting my liquor license.

So I turned both night clubs into strip joints temporarily, while we fought it out in court. The City Council forced my hand in the situation. At the time, it was a last ditch effort to avoid losing everything. But my actions in this situation are not something I'm proud of today.

This transformation worked for a while, but after sinking all my personal savings into the struggling downtown club, I couldn't keep both clubs afloat. Without parking for The Ritz, I was barely breaking even. I wasn't accustomed to failure like this.

An Anchorage Daily News article reported the following, albeit a bit skewed in the details:

> *In 1986, as owner of the now defunct Spenardo DiVinci's in Spenard and The Ritz Showclub on downtown C Street, George LaMoureaux won a faceoff with the Anchorage Assembly, which threatened to try to block the liquor licenses unless he promised not to feature topless dancers. He first promised, then reneged but the assembly failed to follow through on its threat.*
>
> *The family is out of the bar business now, he said, having lost both liquor licenses in bankruptcy proceedings*[11]

The article says that I promised not to feature topless dancers then reneged on that promise. But in fact, my attorney made it clear to the Assembly that I was signing under duress and that in no way was it a binding agreement. Even so, it's one of those situations in which you can win the battle but lose the war. And due to the fact that the mafia was involved, I lost the war. I took no personal losses other than financial.

---

11   Sheila Toomey, "Looking for Cartoons? Lots of Cartoons? Try Channel 14 on Your UHF Dial," Anchorage Daily News, February 19, 1988, p. A1.

Under these circumstances, I was forced to file for Chapter 11 protection for The Ritz and Bogie's. I had tried to support both clubs with Bogie's income, but could not. Consequently, I spent all my personal reserves trying to support both clubs. At last, having drained all my resources, I took the clubs from a Chapter 11 to a Chapter 7 bankruptcy.

Immediately following my bankruptcy, investors who trusted me and knew I would keep my word presented me with the opportunity to take on 25 nightclubs in Florida. These clubs were extremely profitable, each earning a million dollars annually. The investors coaxed me, "George, all you have to do is sign your name here and they're all yours." They had seen my success and really wanted me to operate their clubs. But I didn't want to get back into that lifestyle of drugs, alcohol and bad morals destroying families. So I turned down their offer. I'd never go back into that business again.

So, in a very short time, I had gone from the *high life* to *no life!*

My life was messed up at this time too. I didn't know God's ways and I had negatively impacted others' lives. It wasn't enough that *I* didn't smoke, drink, gamble, or do drugs. I had created an environment that promoted these vices in others. I also had a bad temper and a long string of girlfriends.

Sharon, the woman I had been with for seven years, had a son by me, whom we named Blake, but went by his middle name, Cody. I'm convinced that my nightclub business destroyed my relationship with Sharon. She developed a relationship with a man named Scott, and eventually married him.

Scott was jealous and intimidated by me, fearing that I would come between him and his wife if I came around visiting our son, Cody. So they decided to prevent me from seeing my son, even though I won the right to do so though court proceedings.

Afterwards, they continued to defy the court order that allowed me to be with Cody. Finally, I got tired of fighting with them and backed off. I didn't want Cody hurt in the cross-fire. Today, Cody is a gifted writer and graphic designer. His website displays his bio:

> *Born and raised in Anchorage, Alaska, Cody grew up in the 90's and early 2000's snow and skate culture. A time defined by crews like the Juneau Boys and BTN. Today, that era and those crews stand as an unprecedented snapshot of what was and what still can be.*
>
> *Cody received a degree in journalism from the University of Nevada, Reno. He works as a freelance writer and graphic designer. He has designed logos for local companies and written articles for national publications. He now works as Content Director, Editor, Publisher, Marketing Hustler, and Event Coordinator at Crude.*[12]

I'm very proud of Cody! He's also a world-class snowboarder and skateboarder and has been in many movies. And although through no fault of my own, I mourn the fact that I have no relationship with him.

---

12 http://www.crudemag.com/about.

Chapter Eight
— SHE LOVED CARTOONS —

To hell with circumstances; I create opportunities.
— Bruce Lee

MEANWHILE, THERE I was without a business. I couldn't sit still. I wouldn't be beaten. As I considered various options for a business, the idea of going into television entered my mind. Before I got into the television business, the only knowledge I had of television was intimate product knowledge due to the fact that I watched everything. The television industry interested me, so I just decided to pursue it and become a quick learner.

I've employed this strategy many times over, hopefully gaining some skill at it. When I go after a business about which I know nothing, I read everything I can get my hands on. Back then, I went to local libraries and read all the books and articles I could find on a subject. I use the internet widely today, but those were the days before the World Wide Web.

I also called companies and interviewed their experts. I often got on a plane and visited an organization and analyzed every facet of their business. I would look for ways to improve on what they were doing and implement those improvements. At times, I actually hired private investigators to do some of my fact-finding work for me.

After closing down my clubs and the dust had settled, I had some money left over and was able to raise additional funds from friends who had seen my success in the past. With this investment capital I applied for a television broadcasting license and purchased some equipment from a defunct station. My intent was to launch a small television station in Anchorage. At the time, I owned a five-story house on Hillside and there was enough room on my property to install a very large satellite dish.

Next, I looked into programming options, but there were only limited opportunities for quality networks with whom I could become an affiliate. I considered several existing networks and a few startup networks including Fox that were beginning to broadcast nationwide. But I didn't care for what they were doing, so I decided to go out on my own and do something new and unorthodox within the broadcasting community.

One idea I entertained was to launch an all-movie channel. But I discovered that the way they bundle movies for sale I would have had to purchase a large number of B-movies to obtain a few A-movies. I didn't want to create some second-rate channel, so I continued to investigate other opportunities.

Some years prior to this a friend of mine had started an all-music television station called Catch-22 and I really liked what he had done. About the same time, MTV hit the airwaves. I was looking for something like those ideas that would provide just the right programming and fill a substantial niche.

At the time, I told my fiancée Paula that I was going into the television business. I had observed that she really liked cartoons, because they made her feel good. Paula's love for cartoons and my desire to please her served as the spark that got me to thinking, "What about an all-cartoon channel?" So I began researching the idea. I discovered that cartoons were aired exclusively on Saturday mornings and for a short time on weekday afternoons.

Cable TV was new to Alaska at the time and only broadcast a few channels including: MTV, news, sports, weather, and HBO. It also occurred to me that the newspaper had its various sections: news, sports, weather, and *comics*. The major niche that television was missing was cartoons! Why not create a Cartoon Channel? That was the novel idea for which I had been searching!

I also saw building The Cartoon Channel as a means for expressing my love for Paula and bringing her happiness. Convinced of its inevitable success, I threw myself into launching The Cartoon Channel with every resource and ounce of energy I possessed. But before I continue to unfold the development of The Cartoon Channel, I must say a few words about Paula.

I was deeply in love with Paula and there wasn't anything I wouldn't have done for her. At our engagement, I slipped a two-and-a-half carat diamond on her hand.

Ever and always the entrepreneur, I also decided to help Paula into business by starting a women's boutique that we named Paula's of France. We chose an ideal location for this upscale, posh retail establishment at Alaska's largest mall, Dimond Center (yes, it really is spelled like that) in Anchorage with 130 stores. The boutique overlooked the mall's indoor ice-skating rink.

Unfortunately, sometime later Paula and I had a falling out. However, I still cared greatly for her. Although we tried to rebuild our relationship, we eventually parted. Losing her was very difficult.

I continue to call her every year on December 7 (Pearl Harbor Day) to wish her a happy birthday and hear that she's well. Paula and I remain friends to this day. In the middle of launching The Cartoon Channel, all this drama was occurring in my private life.

After some time, I began dating again and eventually met Terri. Terri was a faithful companion and encouragement to me in all the turmoil that followed.

Chapter Nine

— LAUGHTER AND ADVENTURE COMING YOUR WAY! —

A goal is not always meant to be reached, it often serves simply as something to aim at.
Bruce Lee

INITIALLY, EVERYBODY I shared my idea with for a Cartoon Channel thought I was crazy. But I've learned long ago not to listen to naysayers. In the beginning, I built the entire television station at my house on the Hillside in Anchorage. I put up a broadcast center, the giant satellite dish, and the whole works. I had my television station licensed up there and it was turn-key ready to go.

But soon all the neighbors around me started complaining because I was going to broadcast from my home. At the time, this was a neighborhood in which all the wealthiest individuals in the largest homes lived. In particular, the neighbors didn't like my commercial size satellite dish.

On a day which apparently there wasn't much else to report on, the Anchorage Daily News ran a big article about my satellite dish. I can laugh at it now marveling at how I thought the neighbors would ever put up with that monstrosity of a dish. The paper quoted Jim Barnett, a South Anchorage Assemblyman who was trying to mediate the situation between my neighbors and me. I believe Barnett may have understated the magnitude

> *I first met George and his brother Bart when I started taking Karate from Robert Alejandre. George and I became good friends and often invited me to participate in his business ventures.*
>
> *When George was running The Cartoon Channel, it so happens that I was working for Merrill Lynch and my office was in the same building and across the hall from George.*
>
> *George's Cartoon Channel was the first in the country. Even though I had never been one to watch cartoons, The Cartoon Channel was great and really filled a need!*
>
> *– Tim Whitworth,*
> *VP Investments, UBS Financial Services Inc.*

of the problem when he stated, "The neighbors are quite upset, the thing does dominate the view, and the view is everything up here."[13] Rather than fight with my neighbors, I moved my operation to downtown Anchorage and rebuilt the entire television station on top of the Frontier Building—the tallest building in Alaska. In every situation where I built something, I was the designer and general contractor. I took over a major portion of the penthouse across from Merrill Lynch and from there I launched The Cartoon Channel.

Ironically, I ended up selling the satellite dish, because I decided to use commercial video tape and fiber optics instead. I sold the satellite dish to my friend, Dr. Jerry Prevo of the Anchorage Baptist Temple, where it still stands today. He had a television station and needed a satellite dish so I made him a favorable deal.

While putting The Cartoon Channel together, I immersed myself in a huge learning campaign. I travelled all over the US, Europe and even the Soviet Union to meet anyone and

---

13   Don Hunter, "Neighbors Don't Like Satellite Dish in the View Municipality Says Use for Business Could Violate Zoning Regulations," Anchorage Daily News, March 14, 1987, p. B1.

everyone in the Cartoon industry. Before long my name became associated with those who know and run in that industry.

I began flying to Hollywood to attend all the film conventions, buying cartoons from MGM, Hanna-Barbera, Filmation, and others. The Pink Panther cartoon was my fiancée Paula's favorite and when they told me how much it would cost to broadcast it, I opened my briefcase and astonished them by counting out stacks of cash for the purchase.

> Of all the things George has worked on, The Cartoon Channel impressed me most.
> He almost killed himself working on that. What with setting up satellites in Russia, other countries, meeting with leaders of industry all over the world.
> He didn't go to school for that, that's all him. He just went out and did it!
> – Jerry LaMoureaux, Jr.

On February 14, 1988, The Anchorage Daily News announced the launch of The Cartoon Channel:

> *Boing! Splat! Wheeeeeee! So you tawt you taw a puddy tat, on television, along with Bugs Bunny, a pink panther and green Gumby, all at odd hours of the day and night. Welcome to Channel 14 on the UHF dial, the world's first and only 24hour cartoon channel. Probably. But definitely originating right here in Anchorage.*
>
> *Channel 14 began broadcasting last week from a studio on the top floor of the Frontier Building on 36th Avenue off C Street. The format is simple. Cartoons. All day and all night.*[14]

---

14 Sheila Toomey.

An article from The Anchorage Times that same year gushed about channel 14, The Cartoon Channel, "Channel 14 is not only a first for Anchorage, it's a first for the world."[15] The Cartoon Channel's byline read, "Laughter and Adventure Coming Your Way, 24 Hours a Day."

The following excerpt from The Cartoon Channel Business Plan further demonstrates its success in Alaska:

> *Sonic Cable Television of Alaska/Prime Cable [now GCI Cable] has been carrying The Cartoon Channel on Channel 44 since June 1988. It was added to the channel line-up due to consumer demand. A survey of subscribers conducted by Sonic Cable Television in October 1988 revealed that 87% of The Cartoon Channel viewers said it added value to their cable service. After less than six months on the cable system, a remarkable 48% said they watched The Cartoon Channel and 67% said they watched for an hour or more each day. These are exceptionally positive results, especially for a new television station and basic cable service.*[16]

I launched The Cartoon Channel in Alaska and almost overnight it became the number one watched channel on all the cable networks for all of Alaska. The Cartoon Channel dwarfed everything else! We broadcast The Cartoon Channel via the Alaska satellite, which had a broad reach into all of Alaska, parts of Canada and the Soviet Union, and northwestern United States.

---

15   Ann Chandonnet, "Ehh, What's Up, Doc? Manager Says Channel 14 Offers 'What City Wants.'" The Anchorage Times, Saturday Morning, August 20, 1988, p. A1.

16   The Cartoon Channel Business Plan, 1989, p. 39.

Due to the fact that a portion of the Soviet Union was already enjoying The Cartoon Channel, I decided to market it to their whole country. In 1989 I visited the Soviet Union in order to establish agreements for broadcasting The Cartoon Channel throughout the Soviet Union and engaging their satellites for this purpose. This was still two years prior to the dissolution of the USSR, so I was overwhelmed by the warm reception they gave me.

In Moscow and Magadan I met with all their big wigs: Boris Semyonov—one of Mikhail Gorbachev's right-hand men, the Minister of Finance, The Minister of Culture, The Minister of Communications, and a whole entourage of other important Soviet officials.

Based on recommendations from others who had traveled to the Soviet Union, I had taken a very large case full of watches, candy, rosaries and other miscellaneous items to give away as gifts of appreciation. For years, my father and mother had made rosaries for the Catholic Church and gave me hundreds of these to distribute. Rosaries had been contraband under Communism. But in spite of this, they let me bring it all in through customs. While in the Soviet Union I always wore a suit and overcoat. Girls would line up to meet me as though I were some celebrity and I must confess that I did enjoy their attention.

> *I first started working with George in the late 1980s. What drew me to work with him is that we're both entrepreneurs and have out-of-the-box minds. George has a million ideas a minute. He's very imaginative! George is an excellent strategic and big picture thinker. He initially helped me put together a lease on some oil reserves in Alaska.*
>
> *– Joe Esquivel, CPA, Entrepreneur and CFO of numerous companies*

> *George and I started dating in June, 1989. At the time, I was a single mom with a four-year-old daughter, Tamra.*
>
> *In one of those odd turns of life, we discovered that both George and I had been born in Oakland, CA. Additionally, George's childhood friend Mike Hoffman, was the older brother of my best friend, Kathy Hoffman. So we had likely spent time together in our youth without knowing each other.*
>
> *When we met, George had been very popular in the nightclubs. We enjoyed a pretty exciting dating life. We were very much in love—very passionate. We were together about five years before our daughter Ashley was born and another seven afterwards. George and I are still close today.*
>
> *– Terri Lane*

My hosts took me to fancy restaurants and their beloved steam rooms. I also met with the KGB and the Russian Mafia. There was always someone pitching me something. One item they tried to sell me was footage of the Chernobyl nuclear plant disaster that had occurred in 1986. They thought these films would be worth a lot of money in the US and wanted me to smuggle them back to the States to broadcast them.

On one merry evening, my hosts taunted me in their thick Russian accents, "If you don't smoke or drink you're not a man!" So for the first time I drank some Vodka with them. I was sick for days!

At that time, the Soviet Union was broadcasting on just three television channels, the third of which aired the news for only a couple hours per day. I suggested that we utilize the remaining capacity on this third channel to broadcast The Cartoon Channel. They thought the idea was ingenious and their third channel became The Cartoon Channel.

The more I worked with the Soviets, the more it confirmed that The Cartoon Channel would be everything I intended it to be, the world's largest television channel.

Chapter Ten
— КАРТYН —
(CARTOON IN RUSSIAN)

Cherish your visions and your dreams, as they are the children of your soul; the blueprints of your ultimate achievements.
Napoleon Hill

IN ORDER TO facilitate world-wide broadcasting, we planned to utilize Soviet satellites Gorizont 9, 10 and 11 along with US satellites Satcom 1 and 2. The Soviets were very excited about launching The Cartoon Channel. In view of the levels of cooperation agreed upon by both the Soviets and the United States, this involved quite a monumental undertaking and represented a milestone toward goodwill on the parts of both governments.

Many readers might be thinking about now, "How did we translate all those cartoons?" But in fact, many of the cartoons—especially the classics like the Pink Panther, Road Runner, and Tom and Jerry had no audio script, only action and music. Also, the Soviets not only agreed to allow scripted cartoons to air in English, but viewed this as a good way to further the mastery of English among their people. English speaking cartoons were also supplied with subtitles.

I developed business plans in English for the Soviet Union and then we had every page translated into Russian. These were all signed and stamped with official seals of the Soviet government. I also had my video presentations translated and dubbed into Russian. This was a massive effort! We were launching the largest television channel in the world.

In order to provide the reader with the full magnitude of this enterprise, I've decided to include the Executive Summary from The Cartoon Channel Business Plan here:

Laughter and Adventure Coming Your Way, 24 Hours a Day

A brand new concept in cable television programming has begun. And is already a remarkable success! The proven format has been market-tested and Nielsen rated for more than a year and a half as an advertiser-supported, basic cable network. The Cartoon Channel provides significant satisfaction levels and potential advertising opportunities on both a nationwide and local cable system basis as shown by a national feasibility survey. The Cartoon Channel is committed to providing an outlet for anti-substance-abuse messages. Five percent of total national advertising availabilities will be donated for messages from, for example, "Just Say No" International, the largest and most successful drug abuse prevention program for children.

George had put together The Cartoon Channel. He was stunned that he was getting Nielsen ratings in Alaska! People were zoning out watching cartoons all day long. The primary viewers were not the demographic that one would have predicted. Even real estate companies were advertising on The Cartoon Channel.

George had showed me what was happening in Alaska and told me he wanted to take The Cartoon Channel national. So he brought me on as Chief Financial Officer and I began helping him put together the Business Plan.

– Joe Esquivel, CFO, The Cartoon Channel

The Cartoon Channel presents animated programming in an appealing package designed for all ages. The combination of classic shorts, modern series, feature movies and brand new programming is popular among children, teens, and parents, as well as advertisers ranging from traditional toy manufacturers to realtors to dentists.

The Cartoon Channel's theatrical cartoons, which were made for movie theaters as preview and intermission entertainment, feature such favorites as Bugs Bunny, Daffy Duck, Porky Pig, and The Pink Panther. These adult-oriented shorts, now enjoyed by all ages, are programmed in prime-time and late-night hours as counter programming to typical sitcom and movie fare. The newer high-tech cartoon series, such as Silverhawks, Thundercats, Robotech and adaptations such as Beverly Hills Teens, Police Academy, and Robocop are programmed in the mornings and afternoons for children and parents. In addition to this programming, there are the classic animated series such as The Jetson's, The Flintstone's, Yogi Bear, and The Bugs Bunny Show. The Cartoon Channel will program animated movie classics such as Return to Oz as family movies-of-the-week.

The Cartoon Channel offers the cable operator a low-cost addition to basic cable packages effectively aiding new sales, subscriber retention, community relations, advertising revenues, rate maintenance and syndicated exclusivity headaches.

Cable television service is in 57% of America's television households (Nielsen, November 1989). It is estimated that cable penetration will increase to 65% by the mid-1990s (Paul Kagan, June 1989). In addition to offering retransmission of local broadcast stations, cable television offers customers a selection of additional channels available for additional charges from more than fifty specialty programmers, including such basic standards as ESPN and CNN and such premium movie services as Home Box Office (HBO) and The Disney Channel. For the twelve months ending in May, 1989, Kagan estimates total cable network advertising revenue at $1.2 billion and total license fee revenue at $576 million, and projects a 20% increase over the next year in each.

After nearly three years of research and market-testing, The Cartoon Channel has formulated a seven-year financial projection. At the end of Year 1 of operation The Cartoon Channel will have approximately five million subscribers generating $2.1 million in license fee revenues and $1.7 million in advertising revenue. Operating costs for Year 1 exceed revenues, but projections show by the end of Year 2, an operating income of $1.5 million. Overall debt service (interest and principal) for $14.0 million is accommodated. The Cartoon Channel has entered substantial negotiations for the lease of the Filmation cartoon library of 1,000+ half-hours.

Subscriber estimates are based on historical averages of start-up basic cable services. License fees, ratings and advertising rates are based on industry averages with consideration for both start-up ventures and current trends, including "Madison Avenue's" newly discovered value of cable advertising.

Programming costs are based on extensive negotiations with various cartoon libraries to rent and purchase material. The outright purchase of programming is also being considered. Such purchases would be offset by a reduction in rental fees and would add revenue streams from merchandising and secondary licensing of such purchased material.

In a recent interview in Broadcasting/Cable, Kay Koplovitz, President of USA Network, said its children's programming block has the highest profit margins of any block on the network. The projected success of The Cartoon Channel capitalizes on Koplovitz' remarks. USA Network, along with other cable and broadcast outlets using cartoon programming in selected dayparts represent competition only in the narrowest sense. Whereas a programming service must be well positioned to attract audiences and compete against other program services, The Cartoon Channel's major competitive strength is that it will be **the first and only all-cartoon channel on television**. Being first out of the box was the main competitive advantage of HBO, the first all-movie channel; ESPN, the first all-sports channel; CNN, the first all-news channel; and MTV, the first all-music video channel. Other competitors to these services emerged, some quickly failed, while others achieved modest market shares. But HBO, ESPN, CNN and MTV still dominate their markets largely because they were first with the best programming.

> *The Cartoon Channel will be the first and best all-cartoon channel, and by being first out of the box it will forestall meaningful cartoon competition. Once The Cartoon Channel is launched, advertised, promoted and positioned in consumers' minds as the Cartoon Channel, just like HBO is the movie channel and CNN is the news channel, a new competitor to The Cartoon Channel would not be able to build audiences and attract advertising.*
>
> *Just as there was room for all-news, all-sports, and all-weather channels, even though the networks and independent stations were featuring similar programming, The Cartoon Channel is satisfied that there is ample room for its unique programming —bringing laughter and adventure your way, 24 hours a day!* [17]

I retained Malarkey-Taylor Associates, the largest telecommunications and cable television consulting firm in the world. They were very impressed with what I was doing. I decided to give them a small interest in The Cartoon Channel and hired Charles F. Hookey to serve as The Cartoon Channel's President. Mr. Hookey had been one of Malarkey-Taylor's Associate Vice Presidents, a member of their Board of Directors, and a veteran of the cable television industry. General Robert M. Montague and my brother Max also served on the Board of Directors. I'll say more about General Montague's involvement shortly.

One of the primary factors we had going for us on the likely success of a nationwide Cartoon Channel was that we were "the first out of the box." The Cartoon Channel was the

---

17   The Cartoon Channel Business Plan, 1989, pp. 6-8.

first and only all-cartoon channel on television. Being the first out of the box, we believed would prevent meaningful cartoon competition.[18]

Plans were laid and contracts established to air The Cartoon Channel nationwide via GE Americom's Satcom I-R satellite. The Cartoon Channel Business Plan made our plans clear, "The Cartoon Channel will be a worldwide company."[19] The Business Plan goes on to describe strategies for pursuing markets in Canada, Mexico, and Central and South America.

At the core of success for The Cartoon Channel was an ever expanding library of videotaped cartoons, old and new, leased or purchased, from every cartoon distributor in the world with whom we could do business.[20] So, I continued attending film producing conventions and purchasing more and more cartoons. I became a well-known and respected figure at these studios and among Hollywood celebrities.

In one humorous incident early on, I had flown to Hollywood to meet with Lou Scheimer, founder and former president of Filmation, the second largest cartoon library and animation studio (Hanna Barbara being the largest). I was sitting in the reception area when Mr. Scheimer came out of his office. He scanned the otherwise empty waiting room, looked at my youthful countenance and asked me, "Where's your father?" Mr. Scheimer later graciously agreed to sit on the Board of Directors for The Cartoon Channel.

---

18 The Cartoon Channel Business Plan, 1989, p. 98.
19 The Cartoon Channel Business Plan, 1989, p. 89.
20 Sheila Toomey.

Chapter Eleven
— RIDING THE WAVE —

The reason you went so far and were so successful, George, is because you didn't know you couldn't do it!

Patrick McCourt

LOU SCHEIMER INFORMED me that Paravision had purchased Filmation and suggested that I contact billionaire Michael Stevens,[21,22] the Chairman of Paravision. Filmation was owned by Paravision, which in turn was owned by L'Oréal, a subsidiary of the Nestle Company.

I contacted Michael Stevens who then invited me to meet with him and François Dalle in Paris, France. François Dalle was the former CEO of L'Oréal and largest shareholder of Nestle. I accepted his invitation and immediately went out and purchased seven brand new suits, new shoes, a Rolex watch and a new briefcase so I could attend our meeting in the appropriate attire.

In Paris, I stayed at the Ritz Hotel at my own expense. I spent nearly $50,000 to attend this meeting. I also spent hundreds of thousands of dollars flying my attorney, Ralph Duere, Joe Esquivel, my brother Bart and consultants from Malarky-Taylor to a subsequent meeting in New York. These meetings were all driven by the global launch of The Cartoon Channel.

---

21 http://articles.latimes.com/keyword/michael-w-stevens.
22 http://www.bloomberg.com/research/stocks/private/pers.

In Paris, I personally presented a very sophisticated package describing The Cartoon Channel to Michael Stevens and François Dalle. To this meeting, I brought the Business Plan and Feasibility Study, bound in black leather with gold embossed lettering, and showed a professional, eye-popping video presentation of The Cartoon Channel. Both men were extremely impressed with our business plan and success thus far. As a result, Nestle agreed to take a ten percent interest in The Cartoon Channel and offered to let me purchase the Filmation cartoon library for $25 million.

With that said, the remaining funding would easily become available. Because of Nestle's participation, other investors would flock to grab a share of this investment. Things were looking really good! We were signing up networks and investors and had almost all our funding committed right out of the gate.

After my meetings with Michael Stevens and François Dalle in Paris, they invited me to New York to meet with the law firm Leob & Leob who represented Paravision and Nestle. Leob & Leob is the world's largest entertainment law firm located in the Helmsley Building at 230 Park Avenue, downtown New York City.

For the better part of that year, my team and I worked out of the offices of Leob & Leob. This arrangement came about through the direction of Michael Stevens and François Dalle. All the while, I was living a few blocks away at 38th and 1st at The Corinthian in a 42nd floor condominium. This offered a quick walk to Time's Square and the United Nations Building. Interestingly, this building was featured in the movie *Independence Day*.

I had brought Joe Esquivel onto The Cartoon Channel Team as Chief Financial Officer and he accompanied me to New York. Joe has a brilliant mind for finances and has been a good friend and trusted associate in many of my business ventures since then. Joe is a Wharton School of Business graduate and happened to attend there with Donald Trump. While at Wharton, Joe roomed with Toten Barcardi, the heir to the multi-billion-dollar Barcardi fortune.

Early in his career, Joe was the head of logistics for Arco during construction of the Trans-Alaskan Pipeline. Joe also served as accountant for two Alaska governors: Walter Hickel and William Sheffield. Joe has also functioned as the forensic accountant for F. Lee Bailey, the defense attorney in the O.J. Simpson trial, and as an expert witness in the US court system. Joe is a genius, credited with taking over 50 companies public.

One day I was in my office in New York, when Michael Stevens dropped by and said, "George, I would like you to meet someone." And with that Jacques Cousteau stepped around the corner and into my office! He came over to me, put his arms around me and with his thick French accent said, "I love your Cartoon Channel idea!" Apparently, Michael had been bragging to him about The Cartoon Channel. I mentioned to Mr. Cousteau that I too enjoyed scuba diving and expressed how much I appreciated his work. Jacques Cousteau was an amazing person and I felt honored to meet him.

Travels with The Cartoon Channel brought me into contact with other famous people as well. I often flew to Los Angeles to attend television tradeshows and purchase cartoons. There, on several occasions, I met Don King (the famous boxing promoter). One time, an acquaintance of mine introduced Vanna

White to me. She is a lovely lady and she offered me her phone number, but I was dating Terri at the time, so I declined.

In order to promote a great cause and foster good will for generations to come, I decided to give a portion of all The Cartoon Channel advertising availabilities to the Just Say No™ war on drugs campaign in perpetuity. In order to facilitate this, Malarky-Taylor set up a connection for me with President Ronald and Nancy Reagan. The Just Say No™ war on drugs campaign was the brainchild of Nancy Reagan and she served as the campaign's Honorary Chairman. Brigadier General Robert M. Montague Jr. functioned as Vice President of Development for Just Say No™ International.

In view of our intentions, the First Lady directed General Montague to serve on the Board of Directors for The Cartoon Channel. I flew to Washington DC to meet with General Montague. In a letter to me dated June 12, 1989, General Montague wrote:

> *Congratulations on your progress in setting up The Cartoon Channel on cable television. It's a marvelous idea, which people who know tell me will come to fruition and will be a success.*
>
> *Let's formalize a simple, but significant relationship! We want to assist you actively in completing all the necessary steps to put The Cartoon Channel on cable for American families.*
>
> *Looking forward to the day soon when The Cartoon Channel will be carrying the "Just Say No" message.*[23]

---

23 Robert M. Montague, Jr. Vice President, Just Say No™ International, letter dated June 12, 1989.

As things were really ramping up, Michael Stevens decided to give me a Gulfstream G-550 jet in order to fly around the world to sign up cable affiliates. We planned our rigorous flight schedule and included it in The Cartoon Channel Implementation Plan.

Our financial projections for the national and global launch of The Cartoon Channel reflected very lucrative returns. If The Cartoon Channel's success in the test market was any indication of how well it would perform nationally and globally, then The Cartoon Channel promised to be extremely profitable! Also, because we were broadcasting *recorded* material, the overhead for The Cartoon Channel was very low for a channel with its reach and universal appeal.

All projections about the success of The Cartoon Channel were extremely favorable. We truly believed we were riding on the crest of a very high wave toward success!

Sometimes George's actions were so unorthodox he would embarrass the hell out of me!

One time we were meeting with the top brass of a number of corporations in the Board Room. George had gone out and bought kiddie pencils with cartoon character heads on the erasers. He went around giving one to each participant in the meeting. I wanted to crawl under the table!

But to my surprise, a billionaire in the room was obsessed during the meeting with who had what cartoon character. After the meeting he went around collecting the toy cartoon heads to give to his children! George never ceased to amaze me!

– Joe Esquivel, CFO, The Cartoon Channel

Chapter Twelve
— FROM HERO TO ZERO —

Far better is it to dare mighty things, to win glorious triumphs, even though checkered by failure... than to rank with those poor spirits who neither enjoy nor suffer much, because they live in a gray twilight that knows not victory nor defeat.
Teddy Roosevelt

FOR MORE THAN three years to this point, I had invested every dollar and all my energies into launching The Cartoon Channel—first in Alaska and then taking it global. My travels had taken me from Anchorage to Los Angeles, to Paris, to the Soviet Union and New York. I had worked out of my New York office for nearly a year developing a Feasibility Study, the Business Plan, a 120-Day Implementation Plan, a Delta Phase, and a Protocol of Intentions for the Soviet Union—all combined over 325 pages plus video presentations.

Each of these five books represented countless hours, phone calls, meetings, studies, surveys, number-crunching and projections. Numerous contracts had been negotiated and signed with animated cartoon owners, satellite facilities, cable TV networks, communication products companies, communication services companies, marketing firms, law firms, accounting firms, and many others including the continued daily operation of The Cartoon Channel itself.

As confirmation of the financial viability of The Cartoon Channel, we hired one of the world's largest and most reputable accounting firms, Coopers & Lybrand. On March 31, 1990, the Board of Directors of The Cartoon Channel received a letter from Coopers & Lybrand with their assessment of our finances:

> *In our opinion, the balance sheet referred to above presents fairly, in all material respects, the financial position of The Cartoon Channel, Inc. (a development stage company) as of December 31, 1989, in conformity with generally accepted accounting principles.*[24]

Also, in a letter addressed to me by Malarkey-Taylor Associates on April 15, 1990, they indicated their belief that the Business Plan for The Cartoon Channel "has been obtained from sources considered to be reliable" and that "the basic concept of The Cartoon Channel is sound and that plans and assumptions supporting the financial projections of The Cartoon Channel are reasonable."[25] We were thoroughly and completely prepared.

We were now ready to present The Cartoon Channel Business Plan and attending documents to prospective investment partners in the program. Among others, our attorneys Leob & Leob sent copies of The Cartoon Channel Business Plan along with an elaborate video presentation to billionaire Rupert Murdoch, the Australian and American media mogul; and billionaire Ted Turner, the American media mogul. Of course, each packet of material contained a Confidentiality Agreement.

---

24   Letter from Coopers & Lybrand, Report of Independent Accountants, Anchorage, Alaska, March 31, 1989.
25   Letter from Malarkey-Taylor Associates, Inc., April 15, 1990, included in The Cartoon Channel Business Plan, p. 73.

Sometime after receiving The Cartoon Channel Business Plan, Ted Turner flew to Anchorage to see The Cartoon Channel for himself. Turner met with Marty Robinson, General Manager of Prime Cable of Alaska and one of the executives on my team at The Cartoon Channel, but Turner did not contact or visit me. I didn't find out about Turner's visit to Anchorage until later, but when I did find out about his visit, we all took it as a positive sign.

> Never underestimate a person's ability to rationalize as to why they decide to screw you.
>
> – Joe Esquivel, CFO, The Cartoon Channel

We were now only days away from the national and global launch of The Cartoon Channel. I had flown back to Anchorage to shut down the local Cartoon Channel there in preparation for global broadcasting. On my trip back to Anchorage, I caught wind of a news report in which Ted Turner announced his intent to launch a Cartoon Network. Months later, the following news release hit the press. "On February 18, 1992, Turner Broadcasting System announced plans to launch the Cartoon Network as an outlet for Turner's considerable library of animation."[26] Upon hearing this news, I immediately called Michael Stevens and asked him what was going on. He urged me to get on a plane and head back to New York as quickly as possible.

When I returned to New York, the pieces of the puzzle began coming together. We had made arrangements to purchase cartoons from Paravision (i.e., the Filmation library), MGM, Warner Brothers and Hanna-Barbera. But Nestle (who owned Paravision) became so enamored with the financial prospects of The Cartoon Channel that they purchased Time Warner (a deal

---

26   http://en.wikipedia.org/wiki/Cartoon_Network#cite_note-launch-5.

negotiated by Ted Turner), who owned the Warner Brothers cartoon library. Then Nestle bought out Turner who owned the MGM cartoon collection. Finally, Nestle and Turner contracted with Hanna-Barbera Cartoons exactly as I had specified in my business plan. So Nestle now owned virtually all the major cartoon libraries.

Nestle went around through the "back door" as it were, and cut me out entirely. Ted Turner had apparently been the instigator behind stealing my idea and negotiated the aforementioned purchases and mergers in order to complete the theft of The Cartoon Channel. As Turner unveiled his Cartoon Network, I watched him follow the specific points in *my* Business Plan to the letter.

Suddenly, without warning the wave we had been confidently riding crashed in on us. I was stunned. How could they do this to us?

To add insult to injury, I had also interviewed a former Disney executive as a candidate for an executive position at The Cartoon Channel. She also took my idea back to Disney and they took what I had designed as Toon TV and created Toon Disney.

There was an enormous amount of money at stake here. Our projections estimated that The Cartoon Channel would be

---

George was working on The Cartoon Channel when we first got together. Tamra and I went to New York with him. It was pretty exciting.

But then we got the rug pulled out from under us. We hit rock bottom. We went from having a lot of money to no money. We had to start all over again.

Our relationship back then was definitely based on love and not money, because the money wasn't there!

Back then and today, I most admire George for his perseverance. He's gone through so much. He keeps striving and stays very positive.

– Terri Lane

worth $3 billion within three years. In reality, our projections had been very conservative and the Cartoon Network became worth $30 billion in that amount of time. And it had all just slipped through my fingers.

Turner hired attorney Lisa Murkowski to try to take our trademarks from us claiming we had abandoned them. She was authorized to offer me $100,000 for The Cartoon Channel name, but I told her that unless everyone gets paid, no one gets paid. A lot of investors were engaged in The Cartoon Channel and I couldn't accept this pittance in view of what they and I had invested.

In technical jargon, Turner had *misappropriated proprietary information*. The most common question I get from people when I share my story is, "Why didn't you sue him?" In response, I simply conducted a cost-benefit analysis. I didn't sue Turner, because after assessing the situation this is what I saw.

First, my resources were spent. When I checked what it would cost me to fight this, the price tag would have been upwards of three quarters of a million dollars merely to file the lawsuit because of its magnitude. Even if I had hired the best attorneys money could buy, I don't think they would have had the horsepower to go after Turner with his troop of lawyers and very deep pockets.

Believe me, I fought this for two or three years, but in the end it did not make sense to proceed with a lawsuit. Because of The Cartoon Channel's global scope, it would've become an international lawsuit. A lawsuit of this scale would've gone on for years and my life would have been hell. As Clint Eastwood once said, "Every man has got to know his limitations." I had to walk away from this defeat at some point.

> The Cartoon Channel was one of George's biggest accomplishments. He kept saying it would be the biggest channel that ever existed. George didn't make any mistakes, Ted Turner and Nestle stole it from us. At first they thought we were dumb novices, but when they saw the success of The Cartoon Channel and read our Business Plan they became seriously interested.
>
> We courted Turner and Nestle for two years. We were days away from launching nationwide. We had given up everything to do this.
>
> We showed up one day at the office in New York and we were turned away. The building had a secure reception area and the attendant said, "You'll have to leave or we will call security." This was our office! With their power and clout Turner and Nestle simply waltzed in and took over.
>
> So we went away. I'm not mad. Things come back around. Our other investors must have had hard feelings. It's like something you'd see in a movie.
>
> – Max LaMoureaux

Launching a business has its risks and demands a lot of a man or woman. I expect business to be tough and to cost me something. In building a business I anticipate hurdles, roadblocks, and pitfalls. I know I'll make mistakes, though I make every effort to avoid them. It was one thing to have gone up against the mafia and be taken down by people I already knew were bad. But to have The Cartoon Channel torn from my hands by those to whom I had entrusted my Business Plan was not something I'd prepared for or even considered.

I was devastated. I had lost millions of dollars that I had invested in what would become a multi-billion-dollar network, the world's largest television channel. When all was said and done, I had given many years of my life to The Cartoon Channel. At the time, I had neither the financial, emotional, or spiritual reserves to let go of what I perceived Ted Turner had done to me. But I had to, or it would've paralyzed me and made me bitter. Eventually, God even gave me the ability to forgive Turner.

Louis L'Amour, one of my dad's favorite authors said, "Hatred is an ugly thing, more destructive of the hater than the hated."[27]

But let the record show the following:

Monte Cristo Entertainment Group's Founder, George LaMoureaux is no on the sidelines participant in the entertainment industry. LaMoureaux is the mainstream entertainment media mogul who founded, created and was Chairman and Chief Executive Officer of The Cartoon Channel and Toon TV. These entities are now known as the Cartoon Network and Toon Disney. The Cartoon Network holds the status as the world's largest television network other than CNN in crisis, with 250 to 300 million viewers per day worldwide. Toon Disney is part of one of the world's most recognized entertainment channels.[28]

Purely from an objective viewpoint, Ted Turner is a business man and he took advantage of an opportunity. He grew up with a silver spoon in his mouth, I came from nowhere. The world is full of greed and avarice. As a result, "big fish eat little fish" every day and get away with it.

By the way, I discovered that the Gulfstream jet that had been promised me during the rise of The Cartoon Channel had been given to Arnold Schwarzenegger as a sign-on bonus for his role in *Terminator 2: Judgment Day!* Paravision and Nestle had backed the making of this film and diverted the jet promised me to Arnold. I don't begrudge him the jet. He played no

---

27   Louis L'Amour, *The Lonesome Gods*, (New York: Bantam Books, 1983), p. 306.
28   http://www.montecristoent.com/temp/home.html.

role in the demise of The Cartoon Channel and *Terminator 2* was a great movie.

So, here I was again, "from hero to zero." I closed the television station and sold everything to pay off my debts. For about two years I was very depressed. I didn't want to leave home. I refer to this time as, "one of those Howard Hughes moments." When all was said and done, I had invested nine years of my life working on this project.

Yet, in retrospect, I find it encouraging that it was during my depression that my daughter Ashley was conceived. The loss of The Cartoon Channel and everything associated with it paled in comparison with the joy that Terri and I experienced when Ashley was born. Terri had also brought her daughter, Tamra Esmonde, into our relationship and I helped raise her from the age of four to adulthood. Tamra too was and is a great delight to me.

Even though I had lost this massive project, we *were* successful in launching the world's first all cartoon channel, which ended up being the number one watched channel in the world other than CNN in crisis. In the end, we consoled each other as a family that we were all alive and well and we had each other.

At the beginning of this chapter I quoted Teddy Roosevelt:

Far better is it to dare mighty things, to win glorious triumphs, even though checkered by failure... than to rank with those poor spirits who neither enjoy nor suffer much, because they live in a gray twilight that knows not victory nor defeat.

Even in my depression I refused to "live in that gray twilight that knows neither victory nor defeat." My friend, Colonel

Norman Vaughan, once said, "Dream big and dare to fail." These are words I have lived by my whole life.

I also found comfort in my misery when I was reminded that Walt Disney had experienced similar setbacks in his career. People tend to focus solely on his great achievements in establishing Disney Studios and the Magic Kingdom. But Disney had to file bankruptcy in his early years. Then, Disney and his brother, Roy, and cartoonist Ubbe Iwerks pooled their money and moved to Hollywood where they started the Disney Brothers' Studio.

Their first deal was with New York distributor Margaret Winkler, to distribute their Alice cartoons. They also invented a character called Oswald the Lucky Rabbit, and contracted the shorts at $1,500 each.

A few years later, Disney discovered that Winkler and her husband, Charles Mintz, had stolen the rights to Oswald, along with all of Disney's animators, except for Iwerks.[29]

Like Disney, I too had a great idea and proprietary information stolen from me. And like Disney, I wouldn't give up, but would press on. Perhaps the best was yet to come!

---

29   Biography.com, Walt Disney.

Chapter Thirteen
— SELLING ICE WATER TO ESKIMOS —

My great concern is not whether you have failed, but whether you are content with your failure.
Abraham Lincoln

WE WERE BEATEN down, but not defeated! Looking for a new business opportunity, my brother Max and I launched Cellular World Communications Company. Rather than get a job, I simply launched another business. Many people don't understand this, but I know how to build and manage businesses. So I buy or build and sell businesses as often as some people sell cars. I'm wired to design and launch a business. I don't necessarily want to run a business day in and day out—at least not for long, unless it is a challenging prospect in which my creativity and perseverance would help it achieve some substantial long-term goal. I'm a *pioneer*, not a *farmer*.

Cellular World Communications was the first major telecommunications company in Alaska. Cellular World was the number one place in Alaska to obtain cell phones and service. The company did incredibly well and we sold it for a handsome profit after about a year-and-a-half.

During the Gulf War, when we owned Cellular World Communications, Max and I built one of the largest signs in Alaska at the time. The sign was meant to look like a

> One of my fondest memories as a child is when Dad introduced me to Taco Bell. We had some great times there!
>
> I love my dad. He's funny with great stories, loving, caring, protective, and entertaining. There's never a dull moment with George!
>
> – Tamra, daughter

giant lapel button. This "button" weighed more than three tons and was mounted 60 feet in the air. The "button" was a metal disk 16 feet in diameter with a pipe jutting out of the back sharpened to look like a pin. This giant lapel button featured Saddam Hussein's picture on it within the crosshairs of a gun sight.

The sign served as a combination promotional and political statement. The button said, "So Damn Insane!" as a play on his name, Saddam Hussein. We had contacted Guinness Book of World Records to certify it as the world's largest button, but didn't follow through with it. We received national attention and the press was present when we erected the "button."

In anticipation of selling Cellular World Communications, I was standing in my kitchen wondering and praying about what business to launch next. As I continued to pray and think, I walked over to the sink and filled a glass with water. I held the glass up to the light and looked at it and said, "This is it; water." I sat down at the kitchen table and wrote up the business plan, drew the logos, and came up with the name, Alaska Glacier and then formed Alaska Glacier Beverages, Inc.

At this time, we stepped into the craze of what is known as "New Age" beverages. Alaska Glacier Beverages was the first of its kind to come out of Alaska. Our product slogan was, "So cold, it's hot!" And believe me, it was "hot"! We offered Alaska Glacier in four flavors: Arctic Cherry, Wild Raspberry,

Polar Blackberry and Glacial Wild Berry. Later we offered a non-flavored Natural Sparkling Alaska Glacier Water.

On each attractive bottle, the label proclaimed The Alaska Glacier Story:

> In the early 1600s a majestic, 25,000-year-old glacier was discovered. It was so pure, uniquely vibrant, and virgin fresh, it was named "Eklutna," water between two mountains. From a land so cold and remote there are record temperatures of 83 below zero, and 60 below zero is not uncommon. This is the untouched, untamed land of pure Alaska Glacier, the world's finest water. Exported worldwide from the Last Frontier to you.

> What I admire most about George is that he doesn't know what he doesn't know. He will try anything! He surprises the hell out of me. He has no shame. He gets away with it. He has a very disarming personality.
>
> There are no speed bumps in George's mind, he just moves forward. I wish I could do more of that in my own life.
>
> – Joe Esquivel, CFO, Alaska Glacier Beverages, Inc.

My brothers Max, Jerry and Bart had been and continued to be heavily involved in all my businesses. But with the launch of Alaska Glacier Beverages, Inc., I decided to draw in as many of my family members as possible—they are all talented and bright individuals. Max served as Vice President, Secretary and Director; my dad, Jerry, Sr. functioned as a Director, as did my mom, Mia; my brother Jerry Jr. took on the role of Vice President of Operations. My friend Joe Esquivel followed me to Alaska Glacier Beverages, Inc. as Chief Financial Officer. And finally, we recruited actor Steven Segal as the spokesman for Alaska Glacier Beverages, Inc., whereby he received $1.5 million in stock for his endorsement.

Our water source was genuine glacial water from Lake Eklutna in the Chugach Mountains. With state approval, we built a pumping station at the Eklutna water purification facility at Lake Eklutna. We later proposed building and operating a commercial water bottling plant at the lake, though that never came to fruition.

We bought the full back-cover advertising of internationally circulated beverage trade magazines, including *Beverage Industry* and *Beverage World*. In those advertisements, we offered master distributorships and marketed this program throughout the world.[30] We received over 1,200 requests from distributors worldwide who wanted to carry the brand.

Already by October 1992, we had produced three thousand cases of Alaska Glacier for distributor sampling. And by November of the same year, we had secured the source of Glacier water from the State of Alaska for long-term use.[31]

Almost overnight we were inundated with requests for distributorships. Looking back through the letters from these distributors that we included in our Business Plan, I see letters from 19 states (we were eventually in 30 states), 6 Canadian

> In no time we signed up 52 distributors and Alaska Glacier Water was flying off the shelf! We couldn't supply them all. We could never bottle enough water.
>
> Shipping and distribution was a problem from Alaska. Also, the distributors all bought on 30 days credit, which put a heavy financial burden on us. We only had five-and-a-half turns a year on our money with a 10% profit margin.
>
> Alaska Glacier Water was a great idea, but it ate itself up.
>
> – Joe Esquivel, CFO, Alaska Glacier Water

---

30 Beverage Industry, Vol. 83, No. 11, November 1992.
31 Alaska Glacier Beverages, Inc. Business Plan, TCS Video Production Center, Inc., Production Script, p. 3.

provinces, Ireland, Japan, South Korea, and a distributor with a reach into the Caribbean, South America, and Europe.

In addition to the above, we struck a deal with Saudi Arabia to ship tankers full of fresh, clean, Alaska glacier water to that parched country at a rate of 14.7 million gallons each quarter. Saudi Arabia also planned to purchase cases of our Alaska Glacier bottled water. Brian Crewdson, spokesman for the Anchorage Water and Wastewater Utility, commented on this deal, "We think it's good because we've got the water and we can use the money. We'll be real happy to see this thing go."[32]

In June 1993, we were buying about 24,000 gallons of glacier water each week. We shipped it in sterile stainless steel tanks to a bottling plant in Everett, Washington for processing and bottling.

At that time, I was quoted in The Alaska Journal of Business, "Some 25,000 cases of Alaska Glacier sparkling water products are sold weekly from San Diego to the Bahamas."[33] This was really big business! We had distributors scattered from coast to coast of the United States, we had orders for more than four million cases of Alaska Glacier in 1993.[34] We were now preparing for world distribution.

However, distributors wanted 30, 60 and even 90 days to pay for product orders. Because of this we needed equity participation. We could not pay the interest accruals for up to 90 days on millions of dollars due to the low profit margin in

---

32 Quote from article by Margaret Bauman, "Alaska Firm Plans Glacial Water Sale to Saudi Arabia," The Alaska Journal of Commerce, June 21, 1993, Vol. 17, No. 25, pgs. 1 & 9.

33 Bauman, p. 9.

34 Beth Morris, "Bottling the Essence of Alaska," Anchorage Bypass, Vol. 2, Ed. 15, June 17-23, 1993, p. 2.

selling water. *I didn't want to be in business for the pride of ownership, but for the pride of profit.*

We had secured a number of smaller investors in the beginning, but now needed to go after "elephant" investment capital. These major investors needed to be equity investors rather than merely providing loans. Otherwise we would have found ourselves spending all our profits to service debt. A business can find itself going through all the motions of operation: taking orders, manufacturing product, and shipping product and yet not making any money.

In order to procure more significant funding, we contacted a number of investors including Eklutna Inc., an Alaska Native Corporation, who put up $200 million in real estate assets to be used as collateral to fund Alaska Glacier Beverages. However, the banks refused to facilitate the loan based on their assets due to the fact that the banks had been accused of taking advantage of the Alaska Native Corporation in the past. Though their fears were unfounded, someone quipped that I could be perceived as "selling ice water to the Eskimos."

Actually, we were not having trouble generating interested investors. We already had international promises from Saudi Arabia, and Greece, and others here at home. But these offers all represented very complex transactions that were subject to the details of an agreement. I was always one to try to hold on to as much control of the company as possible, but in retrospect I needed to release more ownership to these potential large investors.

Our search for investors who would put their money where their mouth was, led me to Leslie S. Greyling, the South African corporate raider. Greyling would buy and sell companies, strip them of their assets and go on to raid another company. Greyling contacted me and told me he would like to invest in Alaska Glacier Beverages, Inc. as an equity investor.

Greyling also asked me to help him with some mergers and acquisitions, all of which led to some very interesting involvements.

Chapter Fourteen

— MERGERS, ACQUISITIONS, AND OTHER DEVELOPMENTS —

Ship your grain across the sea; after many days you may receive a return. Invest in seven ventures, yes, in eight; you do not know what disaster may come upon the land.
Ecclesiastes 11:1-2 NIV

LESLIE GREYLING KNEW he could trust me. I did not embrace Greyling's business philosophy and he was always walking the line between what was legal and what was not. So I had to constantly watch myself to make sure we weren't doing anything even remotely illegal or unethical. However, I was good at mergers and acquisitions, and at the time it seemed like a good way both to obtain funds for Alaska Glacier Beverages and earn some extra money.

But in February of that year, Greyling found himself under indictment by a federal court on charges of defrauding investors of millions of dollars in connection with the defunct Winter Park Company. F. Lee Bailey defended Greyling. I was called to the trial in Florida at one point as an expert witness on Greyling's behalf.

I don't know to what extent Greyling had crossed the line and committed a crime. I do know, however, that he had riled up the former governor of Florida, Claude Kirk, who had a personal vendetta against him. Former Governor Kirk used his influence to sic the law on Greyling intending to do him harm.

For a short time during this period I stayed with Greyling in his home, a $25 million mansion formerly owned by John Lennon, in Palm Beach, Florida. The mansion purportedly boasted 27 bathrooms, meaning I could use a different bathroom each day for nearly a month!

The famous architect, Addison Mizner, had originally designed and built this house for himself in 1925. He named the house El Solano, "after the hot, oppressive wind which blows off the Mediterranean Sea in eastern Spain, but also for Solano County, California, his birthplace."[35] The house was later purchased by railroad tycoon, Harold Vanderbilt, and many years later by John Lennon and Yoko Ono. The house stands at 720 S. Ocean Boulevard.

One of the perks of my lifestyle at the time was that I drove nearly any car I wanted. And during this period I was driving a beautiful Rolls Royce.

At Greyling's trial, the jury was unable to reach a decision on the charges, but Greyling pleaded guilty to securities fraud and agreed to be

> George has always been the backbone of the family, the business creator.
>
> He taught me most of what I know and I've added to it. I have one of the largest insurance companies in the state Alaska and one of the highest rated agencies.
>
> He's saved my life with advice several times. He's smart and diligent.
>
> He took care of dad before he died. George lived at his house to take care of him. He never let the family go hungry. He never stops working.
>
> George helped me to be what I am.
>
> – Max LaMoureaux

---

35 http://townofpalmbeachproperty.com/unique-properties-in-palm-beach/

deported to his native South Africa.[36] As a result, Greyling did not follow through on his promise to provide funding for Alaska Glacier Beverages.

However, Greyling's deportation did not affect my ability to work on his behalf. In view of this, I moved to Aruba for a number of months as head of mergers and acquisitions for one of Greyling's companies. One of the acquisitions I negotiated was the Tierra del Sol golf course, the only 18-hole championship golf course on Aruba. I also negotiated the purchase of three other resorts and casinos on Aruba.

On New Year's Eve in 2000, Terri, Tamra, Ashley and I had dinner with F. Lee Bailey, famed O.J. Simpson defense attorney, and his wife at the Tierra del Sol's restaurant where I dined on a regular basis. We enjoyed an amazing display of fireworks that evening. The fireworks were presented on the scale of an intergalactic battle and were some of the most grand and spectacular that I'd ever seen.

For many years until recently I have worked with Greyling. In spite of some of his methods, he had accomplished a lot of great things in the business world. While working for him I traveled all over Europe and South America to negotiate these mergers and acquisitions.

Meanwhile, I continued to secure the right money under the right agreements for Alaska Glacier Beverages, Inc. My search for funds took me throughout the U.S., London, Greece, and South Korea, where I courted other international partners. And my investment bankers traveled to Saudi Arabia on my behalf.

---

36   Tom Stieghorst, "3 Defendants Hold Florida Ties," Sun Sentinel, June 17, 2000.

I had also contacted Michael Stevens again and met with him in New York. I knew that he would be interested in investing in Alaska Glacier Beverages. Michael introduced me to Whale Securities who wanted to invest $50 million in Alaska Glacier Beverages and Spencer Trask Investment Banking made a similar offer. But my board was shortsighted. They thought this was too much money and suggested we conclude transactions with other smaller investors with whom we were already working.

In hindsight, we should have accepted funding from Whale Securities or Spencer Trask Investment Banking. Had I known then what I know now, I would have moved forward with their funding and continued to expand Alaska Glacier Beverages.

Our major challenge with Alaska Glacier Beverages, Inc. was its soaring success. Distributors wanted to place orders in the millions of dollars. But with the manufacturing, shipping, and warehousing and product-placement stages, this all required up to 90 days before we received payment. As a fledgling company, we simply did not have the reserves to function like that. Finally, we reluctantly elected to moth-ball the company.

Babe Ruth once said, "Every strike brings me closer to the next home run." I too had to keep "swinging for the fence." And one way I keep swinging is by having many irons in the fire. My magic formula for making money: Add money and stir!

I had also launched a business in Anchorage in 1995 called V-Tech Motorcycles. The focus of this business was to build custom Harley-Davidson motorcycles, or Super Mod Harleys. My pride and joy was an $84,000 custom Harley. This was an Arlen Ness style motorcycle with a custom Atlas frame and Pat Kennedy 120-spoked wheels on the front and rear. I won awards at numerous events with this stunning all-black-and-chrome show bike. At one show it took the Best Engineered Bike award.

V-Tech Motorcycles was great fun, but not particularly profitable. Even so, it gave me an outlet for my creative talents. V-Tech Motorcycles was a form of therapy for me in creating something I could touch and see without committing to millions or billions of dollars.

Shortly after shutting down Alaska Glacier Beverages, I helped my friend Patrick Moses with his purchase of a Popeyes® Louisiana Kitchen restaurant franchise. Because of my past experience with franchise ownership, I was able to help him with all the documents, structure, and build out of his first location. Patrick continues to run a successful Popeyes® business in Anchorage today.

Over the next ten years, amid the flow of new business ideas and startups, I negotiated the buying and selling of hotels, golf courses, resorts, and many other businesses. I was fortunate to have the God-given

---

George and I have been friends for many years. In the 90s he owed some taxes on a $22 million property that he owned and was short on cash. So he came to me with a unique request.

At the time he owned a beautiful custom Harley-Davidson motorcycle that he had built. It was worth close to $84,000.

George gave me the show bike in exchange for paying his property tax bill. Then he asked that I give him two years to buy the bike back. He said, "If I don't buy the bike back from you in two years, then you can sell it and keep the money. Just don't sell it for two years."

I had that stunning motorcycle displayed in my living room on a plate of glass for nearly two years. Then George came by one day and bought the bike back! He honored his word, but deprived my living room of a beautiful ornament.

– Tim Whitworth, VP Investments, UBS Financial Services Inc.

ability to remember everything that was said and written in negotiations. So when someone tried to pull a fast one on us and change the details of a negotiation, I could say with confidence, "No, we did not agree to that." I was really good at this.

Also around this time, Max and I bought a bail bond franchise called Capital Bonding. Capital Bonding soon became one of the largest bail bonding firms in Alaska.

Max ran Capital Bonding and soon bought me out. The bail bonding business was very profitable, but we were always dealing with alleged criminals and with people who were less than pleasant as clients.

Eventually, Max tired of the bail bond business and I helped him start an all-service insurance agency called Insurance Max. This business is still operating today very successfully under Max's able leadership. Max grew it to become the highest grossing insurance company per capita for Progressive Insurance in the nation. He also represents other carriers. I'm very proud of him!

As the reader doubtlessly recognizes by now, I eat, sleep and breathe business ideas. Couple that trait with the fact that I'm a movie fanatic, and *voila!*—I build a movie-related enterprise. So when the epic film Titanic appeared in 1997 and captured the hearts and imaginations of Americans, I saw an opportunity I couldn't pass up and launched the Titanic Memorabilia Company. I recruited a team of artists and designers to create the art work for these masterpiece memorabilia and reproductions. These are still some of the finest works ever produced to commemorate the Titanic.

We created Triumph and Tragedy paintings, prints, minted coins, earrings, necklaces, and money clips. These were beautiful, elegant keepsakes. We had the endorsement of the Titanic Historical Society (THS) and purchased a number of items from them including the signature of the last surviving member of the Titanic, Millvina Dean.

Ms. Dean was just over two months old when the Titanic sank the night of April 14, 1912. She was one of 706 survivors that night. Her mother was saved, but her father died as the ship sank. Ms. Dean was still alive when the film Titanic was produced, but she passed away at 97 years of age in 2009.[37] Although just a baby when she survived the sinking of the Titanic, we still look to Millvina Dean as a hero. Like Ms. Dean, we all long to rise above whatever challenges and tragedies the tempest of life hurls at us. We soak up strength, encouragement and perseverance from reading about others who struggled, yet overcame their hardships.

---

37 "Last Survivor of the Titanic Dies, Aged 97," Europe on NBC News.com.

Chapter Fifteen
— THE RESORTS —

You just can't beat the person who never gives up.

Babe Ruth

IN THE LATE 1990s, another of the large projects I initiated was the purchase of the Buckner Building in Whittier, Alaska, from Pete Zamarello, the largest property developer in Alaska at the time. The Anchorage Daily News reported, "Whittier's Buckner Building, an abandoned hulk that used to house soldiers, has been bought by a group that hopes to turn the 1950s military building into a resort and condo complex."[38]

The Buckner Building was named after General Simon Buckner, the highest-ranking U.S. military officer to fall in battle during World War II. Bucker was tough, and so is the building that bears his name. Completed in 1953 at the height of the Cold War, this mammoth, bomb-proof structure was intended to house as many as 1,000 soldiers keeping them safe in the event of a Cold War invasion. Whittier, Alaska was chosen for this installation due to its relative proximity to the Soviet Union and the fact that Whittier is an all-year-around open seaport.

The Buckner Building was a 273,660 square-foot facility that boasted a movie theater, bowling alley, rifle range, a post exchange and other amenities. Eventually, the military abandoned the building, leaving it as a ghostly reminder of the Cold War.

---
38 Mike Hinman, "Group Buys Buckner Building," Anchorage Daily News, July 8, 1998, p. G1.

> Pete Zamarello is known in Alaska not only for his extensive property investments and development, but also for an infamous statement he made.
>
> I no longer remember exactly what raised the ire of Pete to insult the local bankers when he referred to them as "pimps and whores."
>
> His slanderous remark evoked a public outcry and he was pressed to issue an apology. Pete consented and apologized to the pimps and whores for associating them with bankers!
>
> – George LaMoureaux

Although Whittier is only about 65 miles southeast of Anchorage, its geography rendered it remote. Whittier was built on the Prince William Sound as a military outpost and port during WWII. In 1943 the U.S. Army Corp of Engineers bored a railway tunnel through the base of 13,300-foot Maynard Mountain providing rail access to the village. The tunnel is named after Anton Anderson, the Army Engineer who oversaw the construction of the tunnel.

In 1998, we got wind of the fact that the State of Alaska was about to begin construction converting this rail-only tunnel into a dual-purpose tunnel accommodating vehicles *and* the railway. Construction began that fall and was completed the summer of 2000, making the Anton Anderson Memorial Tunnel (or Whittier Tunnel) the second longest highway tunnel in North America.[39] This tunnel cost nearly a half billion dollars to build.

The tourist industry estimated that after the dual-purpose tunnel opened, as many as 1.4 million visitors per year would pour into this village of 170 inhabitants.[40] Whittier only had guest rooms for 50 at the time, so the timing appeared perfect for building a resort hotel that would accommodate the droves

---

39 http://en.wikipedia.org/wiki/Anton_Anderson_Memorial_Tunnel#Anton_Anderson_Memorial_Tunnel.
40 Hinman, p. G1.

of tourists that would be gorging this tiny port village. This was an opportunity I couldn't pass up.

I had formed the Prince William Resort Corporation to purchase the Buckner Building. I acquired a Ramada Inn franchise and we began plans to renovate this monstrous military installation into a world class resort complex complete with hotel, condos, mall, spa, theater, shooting range, and convention rooms. We were going to call it the Ramada Prince William Resort, a 465-room resort hotel.

At the same time that I was working on the Buckner Building project—the Ramada Prince William Resort, Leslie Greyling called me about helping him build a Sports World Theme Park in Orlando, Florida. I didn't care for the details of his plan for Florida and told him I thought we'd do better in Las Vegas.

George called me one day and asked if I would come with him to look at the old Buckner Building in Whittier. He was thinking of buying it and turning it into a hotel.

George wanted to walk through the building to assess how he might board it up until construction began. There had been rumors of bears hanging out in the abandoned structure, so we took shotguns with us just in case.

– John Majors, Firefighter, Anchorage Fire Department

I came up with an alternate idea that I believed would work much better. I proposed a $3.4 billion project called The Colosseum and began drafting plans for this colossal structure and operation. I served as the The Colosseum's Project Director and Founder.

The Colosseum will be the world's largest and most versatile sports and entertainment complex. This will be an integrated facility conceived and designed to address an identified need for very large scale sports and entertainment resort facilities in Las Vegas.[41]

With the help of overseas investors, I put money down on 175 prime acres of land in Las Vegas. This land was already zoned for gaming on the "Las Vegas Strip." The property boasted over one-half mile of frontage on Las Vegas Boulevard with easy access via Interstate 15. For comparison sake, Disneyland resides on 168 acres, so this was an enormous piece of prime property strategically located right on the strip. (This property is now right across from the Belz Outlet Mall and next to the Mandalay Bay Casino.)

Even with all the other hotel-casinos in town there was nothing like The Colosseum. "The Colosseum's major competitive strength is that it will be the first and only complex and facility of its type in the world."[42] Also, The Colosseum would meet a need for a large venue sports arena that Las Vegas lacked. To give the reader an idea of the scope of this concept, The Colosseum plan included:

- 6,000 suite resort (50% timeshare, 25% condos, 25% hotel rooms)
- 100,000 seat all-purpose sports stadium
- 1,000,000 square-foot mega mall
- 390,000 square-foot casino
- 630,000 square-foot convention center
- 168,000 square-foot international food court
- Several sport and entertainment theme restaurants

---
41 The Colosseum Business Profile, April, 1998, p. 7.
42 The Colosseum Business Profile, April, 1998, p. 51.

- Sports and Entertainment Hall of Fame
- Huge indoor amusement and water park with Olympic size-plus pools[43]

In addition to the above, we set about to acquire Sports Immortals—the world's largest collection of sports memorabilia and incorporate it into The Colosseum. Our plan was to showcase Sports Immortals at The Colosseum. Sports Immortals would feature the famous collection of Joel Platt with over one million sports mementos. No other collection in the world boasts such a wide variety of authentic memorabilia from so many great athletes.

Part of our strategic financial plan for The Colosseum was to pre-sell: condominium view suites totaling in exess of $2 billion; timeshare view suites totaling in exess of $3 billion; corporate sign advertising and sponsorships in excess of $100 million per year; and pre-lease all of its Mega Mall space earning $139 million per year.[44] Obviously, there were other strong revenue generating elements to this project as well. We estimated that the project "could provide 100% return on investment in 6 years or less."[45]

Our Business Profile touted the colossal benefits of The Colosseum as "The right concept at the right time:"

There is no indoor stadium in Las Vegas that has over 18,500 indoor seats, and for years groups have tried to build a stadium that stands on its own. The

---
43 The Colosseum Business Profile, April, 1998, p. 7.
44 The Colosseum Business Profile, April, 1998, p. 53.
45 The Colosseum Business Profile, April, 1998, p. 2.

essence of their failure may well have been their single source income approach. The key to the success of The Colosseum will be its diversity of revenue streams.

First and foremost are the gaming and timeshare suites. Anyone in the business knows that to have anything without gaming in Las Vegas is like throwing money away and is almost certainly financial suicide for that business. Furthermore, the only business that makes more money than gaming is the timeshare business. Consequently, The Colosseum will have one of the largest casinos in Las Vegas to accommodate the large crowds that patronize the events and occupy the 4486 timeshare and condo view suites which, when sold, will generate over $5 billion. This does not include the revenue streams generated by the remaining hotel suites, the casino and the mall.[46]

I always try to hire the best for any project I pursue, so I hired architect Phillip Thern and Terry Jose to assist me in creating the architectural plans and renderings of The Colosseum. We came up with the designs and then took them to Veldon Simpson to have him confirm the feasibility and development costs.

Veldon Simpson, President of Veldon Simpson-Architect, Inc. We engaged him to conduct a feasibility analysis of the project and he was to become our lead architect. Simpson was the designer of the MGM, Luxor Hotel and Casino, Excalibur and many other world-class resorts. The plans we came up with for The Colosseum were absolutely amazing!

---

46   The Colosseum Business Profile, April, 1998, p. 13.

The architectural design of The Colosseum was a challenge. The goal was to build a facility that had never been built before. The idea was unique: an all-under-one-roof concept—a 100,000 seat stadium with view suites looking into the stadium itself, and additional hotel suites surrounding the entire stadium, along with a design that causes patrons to walk through the mall, the casino and the food court before entering the stadium and convention areas. To this end, the architecture and design team created the first facility of its type, The Colosseum.[47]

In order to provide the reader with a more complete picture of the magnitude of this project, I've Included the Executive Summary for The Colosseum—Casino, Stadium and Resort below.

The World's Most Impressive Sports & Entertainment Complex

The Colosseum is an idea whose time has come. Nothing like it exists in the world today! This multi-faceted complex, through large scale integration of easily matched elements (casino, stadium, resort, mega mall), could provide 100% return on investment in 6 years or less. An idea so appropriate to its time, location and trends that it may well describe an opportunity unequaled in recent history.

The Colosseum's financial keystones are its diversified and synergistic sources of revenue and its mitigated

---

47 The Colosseum Design Profile, April, 1998, p. 2.

risk. The following list of revenue sources illuminate this synergy and diversification:

- *Condominium revenues*
- *Timeshare revenues*
- *Hotel room revenues*
- *Stadium view suite rentals*
- *Mall lease revenues*
- *Casino revenues*
- *Stadium space revenues from sports and entertainment*
- *Convention center revenues*
- *Ticket sales*
- *Television, cable and licensing fees*
- *Signage and sponsorship revenues*
- *Amusement park ticket sales*
- *Food and beverage sales*
- *National restaurant tenant revenues*
- *Hall of Fame ticket and sports memorabilia sales*
- *Sales and marketing of licensed Colosseum promotional goods*

The Colosseum will be located directly on the Las Vegas strip, positioned on the last, large parcel (175 acres) of prime, high profile resort property on the strip. Featuring more than one half mile of strip frontage, its location on the south end of the strip puts it directly across from the soon to be relocated airport car rental, parking, staging and distribution facilities.

McCarran International Airport logged 30 million passengers in 1996 and estimates 40 million passengers by the year 2000. A major portion of Las Vegas visitors rent cars and would have their first, last, largest and most impressive views of the Las Vegas Strip provided by The Colosseum.

Las Vegas is growing by leaps and bounds, with many competitive resort concepts, but The Colosseum concept stands alone.

- *A first class, 5 star resort. It theme and design is futuristic/classic using volumes of glass, polished stainless steel, brass, granite and marble. Replete with high tech statues of gladiators and Olympic sport figures mounted on pillars inside and out. The Colosseum will be the epitome of strength and new technology in every part of its overwhelming design. Each and every part of its design will be massive: doors, windows, pillars, hallways and statues.*

- *The Colosseum structure is an imposing 1740 feet in length and reaches a height of 300 feet providing breathtaking external views for those on the upper floors.*

- *6,000 suites are integrated into the impressive and practical design which surround and provide* spectacular internal views of the stadium and casino areas.

- 3,000 of those 6,000 suites will be sold as timeshares generating revenues of $3 billion, equal to 90% of the entire project cost.

- 1486 suites will be sold as condos; 378 of these are planned as stadium view suites with direct views of the stadium and 1108 suites with casino and elevated stadium views. These condo suites will generate $2 billion in revenue.

- The 1,000,000 square foot mega mall will be the largest in Las Vegas. It will be 3 stories high and will wrap around the entire Colosseum. All traffic will enter through the mall, continuing to the casino and food court before reaching the inner core of stadium, amusement park, convention and entertainment areas. This means substantial foot traffic and a very captive audience that The Colosseum will sell to at every opportunity. There will be moving walkways, and a tram rail system that encircles the entire complex at the third story level of the mall.

- A fabulous, 168,000 square foot international food court provides visitors with easy dining from an amazing variety of menus.

- Arena seating capacity of 100,000+. A variety of configurations will ensure optimum viewing for baseball, football, conventions and other events.

Field level seats will move on tracks and can change from baseball to football mode in 10 hours.

- *Every seat in the house will have front row viewing on an overhead, eight sided video display system.*

- *The 630,000 square foot convention center can be augmented with the 168,000 square feet of the stadium floor to provide almost 798,000 square feet of convention space.*

- *The Colosseum's casino will be one of Las Vegas' largest, at 390,000 square feet, the size of two football fields. All traffic is directed through the casino on entering or exiting the stadium and food court.*

- *The world's largest Sports and Entertainment Hall of Fame will feature the world's largest sports memorabilia collection, displaying and selling memorabilia for virtually every sport known to man. There will also be a world class collection of entertainment memorabilia on display and for sale showing the historical evolution of the entertainment industry and the stars that made it great.*

- *Sports and entertainment theme restaurants will be located throughout the complex.*

- *State-of-the-art fitness facilities, one of the largest in the world, to be used by professional athletes and teams as well as resort guests.*

- *State-of-the-art child care facility will give families an opportunity to experience Las Vegas knowing their children are cared for in a secure environment, making The Colosseum the #1 choice for families.*

- *Movie theaters with wrap-around, home style seating and the latest in audio-visual technology.*

- *A unique billboard system will provide a source of revenue never found in a resort setting. Circling the outside of The Colosseum structure will be a track system that carries 120 billboards 20' by 40' each. At programmed intervals the entire system of 120 billboards will move, constantly cycling and circling the structure. This gives every advertiser the same exposure for the same price.*

*The Colosseum's major strength is that it will be **unique to all of the world.***

Being first will be its competitive edge *in one of the fastest growing industries, the entertainment business, and two of the most profitable industries, casino gaming and timeshare sales. Being the **first integrated large scale casino, stadium, resort, mega mall complex** will discourage like ventures for years to come.*

Just as there was room for the Luxor, Excalibur, MGM Grand, The Mirage, Treasure Island, Monte Carlo and New York, New York, one fact continues to stand out: ***the biggest and most unique still sell out.*** *The Colosseum founders and researchers are satisfied that* ***there is ample room and opportunity for this unique*** *facility, bringing the world's most impressive sports and entertainment complex to Las Vegas.*[48]

Imagine sitting in your living room and watching the Super Bowl live and then walking to other side of your condo and looking down into the casino! This was a massive project of incredible proportions with unprecedented amenities.

So who was going to run this casino? Leslie Greyling had contacted Donald Trump and he had agreed to license and run the casino when we opened it. This would've been the only casino in Las Vegas that Trump ever agreed to run. (Prior to this Greyling had an ongoing relationship with Trump on other projects.) Having Donald at the helm of The Colosseum would be pretty hard to trump!

---

48   The Colosseum Business Profile, April, 1998, pp. 2-5.

Chapter Sixteen
— A NEW LIFE —

For God so loved the world that he gave his one and only Son, that whoever believes in him shall not perish but have eternal life.
John 3:16 NIV

ONE MIGHT THINK that with these two major projects going on—the Buckner Building and The Colosseum, I couldn't possibly have had time for anything else. But there was something going on in my personal life during these years that has impacted my life profoundly.

In 1996, I had returned to Anchorage from London where I had sought international partners for Alaska Glacier Beverages, Inc. In those days, pagers were cutting edge technology. While I was abroad, my service provider had reassigned my pager number to someone else and then gave it back to me on my return to Anchorage. Now I was repeatedly receiving messages from a woman named Shayla Murphy and others for a man named John Majors.

I called and left messages for the many people who had called looking for John Majors, "I'm sorry, but this is no longer John Majors' pager." Shayla was one of those who had called. This went on for some time. Finally one day, I actually spoke with Shayla and during our chat she invited me to a Bible study. At the time, I must admit that I agreed to go to the

> At the end of our relationship George had started going to church. We had had rough periods breaking up, and coming back together.
>
> When George made the decision to follow Christ, he did a total 180! Before, he had been a tough guy, very territorial, jealous, and had few male friends.
>
> Now he's polar opposite—he went from being the owner of nightclubs—behaving like the Godfather—to a caring man.
>
> As a nightclub owner he never had to wait in line anywhere. He was a celebrity, but was always thought to be associated with organized crime—even though he wasn't.
>
> After coming to Christ, George has become compassionate. He has been so drastically changed.
>
> He is a great dad to his daughter Ashley and George has been the only father figure that Tamra had.
>
> – Terri Lane

Bible study for no other reason than she had a cute voice and turned out to be pretty in person too. I attended her Bible study once and found it interesting, but didn't care to go back. However, God was pursuing me.

I should mention here that Terri and I had already broken up. This decision proved to be the best for our relationship as well. We were both on two separate tracks of life and although we had tried many times, we could not bring those tracks together. Terri and I had been together for 12 years. Terri still means a lot to me and we continue to be close friends.

A couple weeks after attending the Bible study with Shayla, I was at the Sears mall in Anchorage when a woman named Gabriella came up behind me and exclaimed, "George, is that you?" My brother Max and I owned a gym in town called Body Tech and I had met Gabriella there. She too was very pretty and she also invited me to go to church with her. So I went with her to the Anchorage Church of Christ. She held my interest, so I continued to go. Coincidently, Shayla attended the same church and we developed friendships between us all. These friendships remain to this day.

Terri and I had been on again, off again. I know my lifestyle was tough on her. I wanted to meet a woman whom I could marry and stay with and I thought church might be a good place to find someone like that. Gabriella and I went out a few times and are still friends.

The Anchorage Church of Christ was meeting in the auditorium of a college. They were a devoted, Bible believing church and focused on teaching people how to study the Bible. The pastor, Matt Hogan, who became a close friend of mine, had given me a number of Scripture passages to read in different sections of the Bible. The more I read, the more interested I became.

Eventually, rather than read random portions here and there, I decided to simply start reading in Genesis and read the Bible from cover to cover. I read the entire Bible over and over again and as a result, my faith increased daily. I also purchased the Bible on CD and have

How has Christ changed George? He was different in a good way. He became humble. When George was younger his life was all about the night clubs, women, money, cars— all a set up to fail.

When George gave his life to Christ none of that mattered any more. He didn't waiver at all. He went 100 percent.

You can see Christ in his eyes. I'm blessed to know him. Most people would lose faith going through what George has but he has grown stronger.

I think better things will happen for him yet. He wants to do great things through Christ. I believe the Lord has a remarkable plan for him.

Most would've given up through the cancer, the loss of so many loved ones, and losing everything he owned. But he found strength in Christ.

My wife was killed in a car accident when my son was two and George stood by me. You will be a better person hanging around George. He will bless you!

– Robert Alejandre, Martial Arts Instructor

listened to the whole Bible repeatedly many times since then and continue to do so today.

A lot of people turn to Christ because they have failed, or experienced some major tragedy and are brought to the end of themselves. But that wasn't my situation at all as I had been very successful. However, I did recognize that I am a sinner and in need of God's mercy, grace and forgiveness. I committed myself to the Lord Jesus Christ, falling to my knees because I recognized who He is. My faith was coupled with both an intellectual and heartfelt decision.

I thought I had trusted Christ as my Savior some years before this, but I had not been living for Christ. I had not known or followed God's Word. Now, as I became more and more familiar with God, His character, and His ways, He began to change me. For instance, I used to have a very bad temper—not a good quality combined with a black belt in Karate. God quenched my temper by flooding my life with His Word.

My most momentous experience with Christ came when I agreed to be baptized. The date was September 20, 1999. My baptism was a turning point for me, because I was publicly declaring that Christ is my Lord and that I am following Him and living for

> *I met George in Anchorage while I was in the military. I was Airborne Infantry in the US Army and had purchased a pager so I could be reached when we were called out on alert.*
>
> *When I transitioned away from that role, I relinquished my pager. Meanwhile, a girl I was dating, Shayla, kept calling my pager and leaving messages. George was the new owner of my old pager number, so he was fielding my calls. That's how we eventually met!*
>
> *I appreciate George most for his persona. When you talk to George he has an uncanny ability to be there totally for you. He connects with others intuitively and even emotionally. That is a rare gift!*
>
> *— John Majors, Firefighter, Anchorage Fire Department*

Him. From my baptism on, I have purposed to live for Christ and according to His Word, the Bible. I enjoy a wonderful relationship with Him. I know He has forgiven me and loves me. When I see Him, my longing is to hear these words from Him, "Well done, my good and faithful servant." (Matthew 25:21)

At about this time, a friend of mine Kevin Creeden opened the Powerhouse Gym in Anchorage and needed help with the marketing. Kevin is a world-class fitness trainer and national and international champion body builder. In order to assist Kevin, I launched Sales Marketing International (SMI) to sell memberships for his gym. Max and I sold 3,000 memberships for Powerhouse Gym in six months. All the while, I continued pursuing and operating my other businesses.

The wheels turn slowly on business projects of the magnitude of The Colosseum and the Buckner Building, but we kept moving things forward. I felt like my personal life was in order and that both of these ventures would succeed. But then, the unthinkable happened…

Chapter Seventeen

— A LOFTY VISION —

Freedom itself was attacked this morning by a faceless coward, and freedom will be defended.
George W. Bush

NINE-ELEVEN, A NUMBER—A date—seared into the hearts and minds of Americans, changed all of our lives. Images still haunt our memories of the passenger jets careening into the Twin Towers, people jumping to their deaths from the burning buildings, mass destruction, grief, sorrow, a sense of defilement, and the irrationality of it all. As a nation we suddenly felt vulnerable and helpless.

The shock waves of the terrorist attacks on the United States on September 11, 2001, continued to rock other structures including businesses and families for a long time afterwards. Overnight, funding for both my projects: the Buckner Building in Whittier, Alaska and The Colosseum in Las Vegas, fell away as financing for the travel and resort industry dried up nationwide. We experienced a sudden, catastrophic shift in national priorities and bankers pulled back all their funding. The fact that the Buckner Building was in Alaska and that the Colosseum was such an enormous project, also discouraged bankers from funding them.

Soon thereafter, without funding for the project, I let go of the 175 acres in Las Vegas. But as with much of our country,

> *George is one of the most passionate men I've ever known. George is always a guy that epitomizes the cup half full.*
>
> *He's got an amazing ability to see beyond current adversity and circumstances.*
>
> *God has given him the ability to lift his eyes to the Lord. I've seen him weep before the Lord. His optimism is incredible.*
>
> *– Pastor Karl Clauson, Radio Show Host and Speaker*
>
> *(I met Karl when he was a pastor at ChangePoint, the largest church in Alaska. Karl was on the cutting edge of bringing the Gospel to Alaska. He is a man of vision and a gifted evangelist.)*

I was still too stunned by this cowardly, hate-filled attack on innocent people—and on decency itself—to focus on the pettier issues of business.

My heart groaned especially for the children made orphans and the children who went missing in this vicious act of cowardice. I realized how fortunate I was to have my daughters Tamra and Ashley safe and sound. I began asking around with the intent to create a fund to help support those children and their families. But I wasn't alone in my compassion and the American people rose up and flooded the Red Cross and other agencies with money and time to help support the victims of this cruel crime.

I told my friend Tom Madden about my desire to help these kids and their families. Tom is author of the book, *Spin Man: The Topsy-Turvy World of Public Relations,* a tell-all tale of the American public relations industry. Before starting TransMedia Public Relations firm, Madden was director of public relations at ABC television and vice president at NBC television.

Tom urged me to consider directing my attention toward an organization that serves children and their families and that would be around long after the dust settled at Ground Zero. Tom told me about Sherry Friedlander in Florida who had founded a non-profit organization called A Child Is Missing.

Sherry started A Child Is Missing (ACIM) in 1997. She created the organization because no community-based program existed for locating missing children, the disabled and elderly (often with Alzheimer's) during the crucial first hours of their disappearance.[49]

Ms. Friedlander has been involved with publishing, public relations and marketing in South Florida since 1964. She started her first public relations firm, Friedlander and Associates, in 1969. Under her leadership, the company grew into one of the most prestigious public relations firms in the area. In 1974, the firm merged to become Cameron-Friedlander Public Relations and Advertising. Again, with her direction, the company continued to expand, attracting not only local, but also national clients.

In 1976 Ms. Friedlander established Lauderdale Publishing and in 1982 began producing business publications, including Broward Economic Development's Guide Book (now known as Broward Alliance), Business in Broward, Art Spirit! and several other publications. She established A Child Is Missing in January 1997, a nonprofit Florida corporation created to assist law enforcement in the recovery of missing children, the elderly and the disabled.

Sherry is well known for her extensive participation in the community. Her community involvement has included membership in The National Association of Women Business Owners, the Cystic Fibrosis Foundation, Delegate to the White House Conference

---

49 AChildIsMissing.org/about

on Small Business, Broward Alliance, and Nova Southeastern University Board of Governors for the School of Business and Entrepreneurship.

Among many awards Ms. Friedlander has received in recognition of her community service are the Kiwanian of the Year award presented in October 2000 by the Fort Lauderdale Kiwanis Club, an award from Women In Distress for 25 years of Distinguished Service, and the prestigious J. Edgar Hoover Award for Distinguished Public Service, conferred in October 2000 by the National Association of Chiefs of Police (NACP), presented by NACP Executive Director Donna Shepard at the Police Museum and Hall of Fame in Miami, Florida.[50]

When a child goes missing, someone dials 911 to alert the local police or sheriff's department. Law enforcement then initiates ACIM's high tech tool which generates 1,000 calls per minute to the neighborhoods and vicinity where the child was last seen. In this way thousands of eyes begin looking for the child in the most likely areas. As of May 14, 2014, ACIM has assisted in the safe recovery of 1,480 children.[51]

ACIM was introduced prior to the Amber Alert program and complements the Amber Alert program by working directly with law enforcement in the moment a lost child is reported missing. ACIM is a proactive system in which they geo-map the last sighting of the child and generate 1,000 calls per minute with a message similar to this:

---

50 http://climbforamericaschildren.com/promoteam.html
51 AChildIsMissing.org

This is the Anchorage Police Department. A child is missing in your area. The child's life may be at risk. Here is the description of the child. Please go outside your home or business right now and see if this child is nearby and call 911 if you see the child.

In contrast, the Amber Alert program operates on the basis of using broadcast channels on television, radio and electronic billboards. The first few critical hours after a child goes missing are the most crucial. That's why ACIM's system is so vital and effective.

I called Sherry and we talked. She was very excited about ACIM's work, but revealed that to date the program only existed in Florida and Rhode Island. I told her I thought we needed ACIM in Alaska and she agreed. So I brought her and her husband up in March 2002. I introduced her to some politicians and the police department and Alaska became the next state to adopt the ACIM system.

I immediately saw the potential for ACIM to expand to all 50 states and my mind's wheels began turning. Just a

One day out of the clear blue, I got a call from a guy named George LaMoureaux. He said, "I think we should have A Child is Missing program in Alaska." I said, "Fine, I do too."

When someone calls me like that I never know what their intentions or follow-through will be like. George was instrumental in showing me that expansion of the program was possible. Alaska became the third state behind Florida and Rhode Island to implement the ACIM program.

George helped introduce us to politicians on The Hill. Teaming up with George was a good marriage.

He helped raise money to continue the work nationwide and we're in all 50 states today.

George always puts a bright light in my eye. He's a great thinker. And we've remained great friends.

He's always there. He was very, very helpful in our expansion.

– Sherry Friedlander, President and Founder of A Child is Missing

> When George contacted me and told me he wanted to climb O'Malley Peak as a fund raiser, my friend, Eric Oakes and I were a bit concerned. Even though George had grown up in Alaska, he was not an "outdoor cat" at all.
>
> I had been an Airborne Ranger and Eric was former Marines Reconnaissance. We asked George, "Are you sure you want to do this?"
>
> Although we were skeptical, we coached George and he did great! We summited O'Malley Peak in the winter and hung out on the summit for a while before coming down.
>
> I think maybe George got bit by a mountain spider on that trip, because he went on to climb Denali and Everest.
>
> – John Majors, Firefighter, Anchorage Fire Department

few months prior to this, I had planned and conducted a fund-raising climb-a-thon for Greatland Christian Church and thought that a similar event would work to promote and raise funds for ACIM.

That original climb-a-thon for Greatland Christian Church took place on October 27, 2001. My friends, John Majors (the former owner of my pager number and now a friend), Eric Oakes and I climbed Mt. O'Malley at 20-40 degrees below zero to raise funds for our church. Eric later went on to climb Mt. McKinley with me. On O'Malley, we did not climb with ropes or crampons, but simply kick-stepped up the mountain using an ice axe to self-arrest. I did take a slider at one point that could've been serious, but was able to arrest my fall with the ice axe.

That fund-raising effort was successful and I thought, "What if we planned a much grander climb-a-thon to help raise support and awareness for ACIM?" I began putting together such a plan to climb Mt. McKinley in the summer of 2002 to benefit ACIM.

Not being a mountain climber, I had no idea at the time what such a climb would require of me, or how dangerous it would be.

Chapter Eighteen

— THE COLDEST MOUNTAIN ON EARTH —

Dream big and dare to fail!
Colonel Norman D. Vaughan

WITH ITS SUMMIT at 20,320 feet, Mt. McKinley is the tallest mountain in North America. McKinley is also known as "Denali" – Athabascan for "the high one"—and is most commonly referred to by its Native name in Alaska. Denali has been dubbed, "The coldest mountain on earth." Even in the summer, temperatures average between 20 and 40 degrees below zero and winds of over 100 mph are common. Winter temperatures have been recorded as low as minus 148 degrees Fahrenheit. (For the reader who would like to learn more about Mt. McKinley, there is a book by the title *Minus 148 Degrees: First Winter Ascent of Mount McKinley* by Art Davidson.)

Although significantly shorter than Mt. Everest (29,035 feet), Denali boasts the tallest vertical relief of any mountain on the planet (18,000 feet). Seasoned climbers quip that "a cold day on Mt. Everest is a tropical (warm) day on Denali." Because of its altitude and location, Denali creates its own climate, constantly building and sloughing its colossal loads of snow and ice.

I must explain at this point, that I had absolutely no experience mountain climbing, hiking or outdoor camping. My idea of a camping trip is a stay at a Hyatt Regency on Maui or

Waikiki. Climbing the north face of Mt. O'Malley at 5,150 feet a few months earlier had been my first and only mountain climb. While potentially dangerous, O'Malley is nothing compared with Denali. But I thought we needed to do something big enough and difficult enough to garner public attention for ACIM.

I contracted Gary Bocarde with Mountain Trip guiding service and he assigned Les Lloyd to lead our Denali climb. Les was a seasoned guide on Denali with a good safety record. This would be Les' 21st expedition to McKinley, having sumittted 13 times. Les and I brought along two assistant guides as well, Richard Baranow and Carl Oswald.

I also obtained the endorsement of the following individuals for the Climb for America's Children event to climb Mt. McKinley: President George W. Bush, Colonel Norman Vaughan, a member of the first Byrd Antarctic Expedition in 1928-1930, Governor Tony Knowles, Congressman Mark Foley, Former Alaska Governor William Sheffield, Former Alaska Governor Walter Hickel, Florida Governor Jeb Bush, the American Mountain Guides Association, and the Mountaineering Club of Alaska. (The reader may view these letters of endorsement on my website: www.climbforamericaschildren.com and enclosed herein.)

I also recruited four other climbers besides me, including Dr. Ginger Southall, Eric Oakes, DeAnna Capps and Patrick McIntyre our videographer. I raised the money for the entire expedition and supplemented what I didn't raise with my own funds. After researching the Denali climb, I put together the following overview and presented it as a pep-talk to the climbing team:

There is no such thing as a walk up route on Denali. We will be carrying heavy loads and dragging massive

> sleds daily. Sleeping and eating conditions will not always be ideal and severe storms may keep us tent-bound for days. We must prepare ourselves mentally and physically for the many challenges of the climb. At any point in time you can make a step in the wrong direction while on a ridge roped and harnessed in with a 2,500 ft. vertical drop on one side and a 4,500 ft. vertical sure death drop on the other side, or step into a glacier crevasse and have tons of ice crunch you like an egg. Furthermore, you could get caught in an avalanche with hundreds of tons of snow and ice turning you into frozen scrambled eggs, need we say more.
>
> We as climbers are risking our lives to save children's lives, do not be confused, every year people die on Denali. It has been said that if you are climbing Denali, get your affairs in order first. People working around the mountain offer to buy your life insurance in order to cash in if you don't come back, the odds are better than the lottery. You see, if the fall doesn't kill you the cold, snow, ice, below zero temperatures or altitude sickness will.[52]

Then, just one month before our climb, I came down with double-pneumonia, code blue, and was admitted to Providence Hospital. I had been training hard by regularly climbing Flattop Mountain and suspect that I contracted pneumonia from a man I ran into on the mountain. On the way down Flattop, I spoke with this man and could tell that he was sick.

When I finally went to the hospital I could barely walk. The doctors warned me that had I not come in when I did,

---

52 http://www.climbforamericaschildren.com/denali.html

> The reader might wonder why George and not Les Lloyd or another experienced climber gave this pep-talk. But this is typical of George no matter what he pursues.
>
> Even though he may have been a novice at the time he hatched an idea, he would study it so thoroughly that by the time the idea was ready for implementation, George was an expert!
>
> – Rob Fischer, Ghostwriter

the pneumonia may have been fatal. For days I lay in a hospital bed with IVs in both arms and doped up with medication. Fortunately, I rebounded and was able to make the climb on schedule after all.

I had lived in Alaska many years and had seen McKinley from a distance, but I had never beheld its full, impressive stature until the day before we began our climb in June 2002. As a climbing team we were on our way to Talkeetna when we had crested a hill and a clear view of the mountain presented itself. There was a pullout along the road so we stopped and got out of the car. Looking up at the top of Denali we could see the clouds scudding by its peak in the jet stream. I remember thinking to myself, "Holy s—-! What have I gotten myself into?!"

Denali is a very physically taxing climb that demands peak aerobic and strength conditioning. My long-time friend and champion body builder, Kevin Creeden agreed to help whip me into shape. I was also still working out regularly in Karate and weight training, but I needed aerobic fitness as well. However, nothing totally prepares you for what the actual climb demands.

When I started the climb, I had bulked up to 200 pounds and was bench-pressing as much as 540 pounds. (I never took drugs or steroids.) On Denali there are no Sherpas or Yaks to assist with the climb. We carried everything ourselves. This meant that each of us carried a large back pack *and* pulled a sled with a combined weight of up to 200 pounds, depending on the size and strength of the climber.

We chose to ascend Denali on the West Buttress route, which is the preferred route of most climbers.

> Nowhere in the world does one travel with so much gear over so much vertical in such a hostile environment. Many stretches of "The Butt" leave very little margin for error (the lower glacier in warm conditions, Windy Corner, the Autobahn, Denali Pass, and the Summit Ridge). Furthermore, the West Buttress is just as exposed as any other route to Denali's legendary weather. Prospective climbers should be highly competent in travel on moderately steep snow/ice slopes and exposed traverses.
>
> The most popular camps are located at 7,200 ft. (base camp); 7,800 ft.; 9,500 ft.; 11,000 ft.; 14,200 ft.; and 17,200 ft. Other camps are located at 12,500 ft. and 16,000 ft., but should only be used under ideal weather conditions as the 12,500 ft. camp is vulnerable

---

I was working at Alaska Mountaineering and Hiking (AMH) when I first met George. He had just climbed O'Malley Peak and told me he was planning to climb Denali for America's Missing Children. George asked me to guide for him, but I declined. I had guided on Denali before and I didn't want to be on the mountain with a bunch of novices. It's too dangerous!

But sometime later I met George's whole climbing team and I saw that he was taking two beautiful women on the climb. Ginger, who was exceptionally gorgeous, very intelligent and well-spoken, talked me into coming with them.

Denali was George's first real mountain climbing adventure. He was extremely fit—muscular, but bulk does not necessarily translate to being a good mountaineer.

But the team was well-equipped and in my opinion they had George's fitness and dogged determination to thank for their success.

– Richard Baranow, former mountaineering guide

to avalanches and the 16,000 ft. camp is very exposed to high winds. The 11,000 ft. camp also experiences avalanches and serac fall, and care should be taken to avoid these two hazards when setting up camp. Above 14,200 ft., snow caves or igloos are usually constructed as a back-up shelter in case bad weather moves in. Total horizontal length of the West Buttress route is approximately 13 miles with about 13,500 ft. of vertical gain. Between base camp and 11,000 ft., the route is relatively flat and the main hazards are crevasse falls. Above 11,000 ft., the route steepens to moderate slopes (35-45 degrees) alternating with flat benches and bowls.

Equipment and supplies are typically carried by sled to 11,000 ft. or all the way to 14,000 ft. Above 11,000 ft., gear and food can be ferried between camps in two trips. West Buttress expeditions average around 17 days, but climbers should take at least three weeks of supplies. A two-to-three day supply of food and fuel should be left at base camp in case weather prevents planes from landing on the glacier (climbers have been stranded for as long as two weeks due to inclement weather).[53]

From start to finish, we planned to be on the mountain for about seventeen days with some buffer built into the schedule in case of foul weather. Woody Allen once said, "If you want to make God laugh, tell him about your plans." In view of what soon transpired on the mountain, God really must have been having a chuckle over our plans!

---

53 http://www.summitpost.org/denali-mount-mckinley/150199#chapter_12

Chapter Nineteen
— "THE HIGH ONE" —

It is not the mountain we conquer but ourselves.

Sir Edmund Hillary

WE FLEW BY small plane to 7,200 feet on the Southeast Fork of the Kahiltna Glacier where the West Buttress route begins. With a built-in reserve in case of foul weather or emergencies, we packed in food for 17 days. Mountain Trip's website adds the following description of this route:

> *The West Buttress of Denali is the classic mountaineering objective in North America. First pioneered in 1950 by the indefatigable Bradford Washburn, it has become the route of choice for most Denali climbers today due to its relative ease of access in this modern age of Air Taxis.*
>
> *The West Buttress route begins at 7,200 feet on the Southeast Fork of the Kahiltna Glacier. It follows the Kahiltna north before ascending up onto the West Buttress proper. Climbers will use a variety of mountaineering techniques to make their way around crevasses and up moderately steep terrain. The route culminates on summit day by following an incredible knife edged ridge to the highest point in North America.*
>
> *Denali is a place of superlatives. Carrying the heaviest pack of your life in the thin air of altitude at such high*

latitudes can make the West Buttress a very physically challenging climb. Extreme winds, heavy snowfall and arctic cold all conspire to make it a serious undertaking. Aspiring West Buttress climbers need to be in top physical shape and prepared to suffer with a smile.[54]

On about the third day up the West Buttress route, one of our climbing party, DeAnna, ran into trouble. The harness with which DeAnna was pulling her sled had begun digging into her sides causing some open sores on her hips. In addition to that, her new boots were breaking in her feet causing severe blisters on her heels. As a result of these injuries and to the great disappointment of us all, DeAnna had to turn back. Two assistant guides took her back to the drop off point while we waited for them to return.

Due to the fact that Denali sits at a very high latitude (about 63° N), the air is even thinner at altitude there than on mountains of equivalent height nearer the equator. At about 10,000 feet, the altitude brought on a massive headache for me. I refused to take altitude sickness drugs and instead combatted the effects of high altitude by guzzling all the water I could drink. Altitude, extreme exertion, cold, wind and sun all unite to dehydrate a person quickly.

People frequently die attempting to climb Denali. Therefore, the National Park Service, who manages mountaineering activities on Denali, has contracted with the Division of Emergency Medicine, University of Utah to study the fatalities with the goal of reducing the number of deaths that occur. In a 2008 report, the Division of Emergency Medicine explained:

---

54 http://mountaintrip.com/alaska/denali/west-buttress/

We retrospectively reviewed the fatalities on Denali from 1903 to 2006 to assist the NPS, medical personnel, and mountaineers improve safety and reduce fatalities on the mountain. Historical records and the NPS climber database were reviewed. Demographics, mechanisms, and circumstances surrounding each fatality were examined. Fatality rates and odds ratios for country of origin were calculated.

From 1903 through the end of the 2006 climbing season, 96 individuals died on Denali. The fatality rate is declining and is 3.08/1,000 summit attempts. Of the 96 deaths, 92% were male, 51% occurred on the West Buttress route, and 45% were due to injuries sustained from falls. Sixty-one percent occurred on the descent and the largest number of deaths in 1 year occurred in 1992. Climbers from Asia had the highest odds of dying on the mountain.

Fatalities were decreased by 53% after a NPS registration system was established in 1995. Although

From the moment I saw Deanna on the mountain, I knew she was out of her element. She no doubt would've done great on a beach in Florida, but not on Denali with crampons and an ice axe!

About the third day of our climb, Deanna recognized she was in way over her head. So Carl Oswald and I escorted her back to the pickup point and put her on a plane.

On our way back up to meet the rest of the party, we were in a storm with poor visibility and Carl fell into a crevasse. He slipped right through the snow into the opening below and fell up to his armpits with his feet dangling below in open space!

I quickly checked my position to avoid the crevasse and stepped over to help Carl up. This was a solemn reminder of the dangers on this mountain!

– Richard Baranow,
Mountaineering Guide

> *mountaineering remains a high-risk activity, safety on Denali is improving. Certain groups have a significantly higher chance of dying. Registration systems and screening methods provide ways to target at-risk groups and improve safety on high altitude mountains such as Denali.*[55]

We experienced avalanches on the mountain virtually every hour of the day bringing down incalculable tons of snow and ice. At night in our tents we could hear the rumble and great crashes of these cataclysmic cascades.

On Denali we encountered no rock—only snow and ice. As we ascended, Les pointed out various landmarks named according to the nationality of climbers who lost their lives there. For instance, the *Orient Express* marked the grade on which a party of Asian climbers slid to their deaths. On another steep slope—*the Autobahn*—a party of Germans perished. On one day of our climb, a tangle of gear careened down the mountain past us. These were sobering reminders of the dangers to which we had submitted ourselves.

About two weeks into our climb, at approximately 17,200 feet, the weather grew increasingly worse until we were forced to set up camp and wait out the storm. Winds of over 100 mph blasted and whipped our tents. The wind chill plummeted to minus 100 degrees Fahrenheit. We were pinned down for over a week.

During that week, we occasionally left the tents to check on each other, or simply move around. On one of those brief excursions outside the tent, I had a remarkable encounter. I met a lone climber in a bright red jumpsuit. He was a very tall and

---

55   http://www.ncbi.nlm.nih.gov/pubmed/18331224

powerful looking man. Shouting to make himself heard over the wind, he told me his name was Gabriel. We spoke for a few minutes and then he said, "George, if you can do this, you can do anything." I had never met this man before. And then he left and I never saw him again. I'll leave you to your own conclusions about that experience. But his words and the fact that he was out there all alone struck me profoundly.

In a situation like we were in, clinging to life enclosed in a puny fabric tent on the side of a killer mountain with the storm raging, I reached for encouragement and vision in all the places I knew I could find it. During that week hunkered down in the tent, I directed my mind back to that great speech that President John F. Kennedy gave at Rice Stadium in Houston, Texas, about shooting for the moon. I've included excerpts from his speech here. No doubt you will see why it energized me:

William Bradford, speaking in 1630 of the founding of the Plymouth Bay Colony, said that all great and honorable actions are accompanied with great difficulties, and both must be enterprised and overcome with answerable courage.

The exploration of space will go ahead, whether we join in it or not, and it is one of the great adventures of all time. No nation which expects to be the leader of other nations can expect to stay behind in the race for space. We mean to be a part of it—we mean to lead it. We intend to be first.

We choose to go to the moon. We choose to go to the moon in this decade and do other things, not because they are easy, but because they are hard, because that goal will serve to organize and measure the best of our

> energies and skills, because that challenge is one that we are willing to accept, one we are unwilling to postpone, and one which we intend to win.
>
> We have given this program a high national priority— even though I realize that this is in some measure an act of faith and vision, for we do not now know what benefits await us.
>
> Many years ago the great British explorer George Mallory, who was to die on Mount Everest, was asked why did he want to climb it. He said, "Because it is there."
>
> Well, space is there, and we're going to climb it. And, therefore, as we set sail we ask God's blessing on the most hazardous and dangerous and greatest adventure on which man has ever embarked.[56]

As the reader will see later, President Kennedy's speech had a profound multi-dimensional impact on my life.

A week of inactivity, extreme conditions, cramped quarters and the lack of sufficient food intensified emotions and spawned lengthy discussions about abandoning the climb. Sometimes in such situations one balances precariously between wisdom and folly; courage and stupidity and it becomes difficult to distinguish which is which. The bulk of the team wanted to go home and tried desperately to talk me into leaving with them. I, on the other hand, told them quite forcefully, "You can all go home, but I'm staying until I summit!"

In the movie *Braveheart*, Mel Gibson's character declares, "Every man dies, but not every man really lives." After all I

---

[56] John F. Kennedy, "Moon Speech", Rice Stadium, Houston, TX, September 12, 1962. (http://er.jsc.nasa.gov/seh/ricetalk.htm)

had put into this I had to see it to the finish! I want to be that man that really lives!

Our discussion as a team was very heated. I was determined to summit. I was not going to run home like a coward. After much debate, we reached an agreement. When the storm finally subsided, half of the team decided to turn back, while the other four of us stayed to summit. Those who left began their descent the very next day before the rest of us summited. That we could attempt a bid for the summit the very next day came at much of a surprise to all of us.

The day we went for the summit, we carried only food and emergency gear. In order to reach the summit of Denali, one must balance across a knife-edge ridge with drops of 2,500-feet on one side and 4,500-feet on the other. As a child, I had battled a mortal fear of heights, but on summit day I walked across that ridge without hesitation. All four of us: Ginger, Eric, Les and I summited.

It took us a day to summit from our camp at 17,200 feet and return to that camp for the night. Our descent to our pickup point required another two days from this camp. During our descent a man climbing down behind us fell to his death, punctuating the stark realities of the inherent dangers of this mountain.[57]

By the time we were making our descent, we had nearly exhausted all our food rations. One of our last meals consisted of dehydrated spaghetti. Eric, with his six-foot-six stature was always hungry. He and I were sharing this meager meal. On climbs like this we employed a very simple arrangement as far as eating utensils were concerned. Our eating utensils for

---

57 See Anchorage Daily News article by Nicole Tsong, "Mount McKinley Claims First Death Since 1998," July 1, 2002.

all meals consisted of a melamine cup with a handle to which a cord was tied to a spoon. This setup kept eating simple and prevented the spoon from wandering off.

Eric and I were scarfing down our spaghetti when I inadvertently dropped my cup which spilled out its contents onto the snow. In one fluid movement, Eric did a head dive with his spoon extended, scooped up my fallen spaghetti and shoveled it into his mouth! I laughed for ten minutes, tears streaming from my eyes!

Both halves of our team completed the descent without serious incident. We had brought food for 17 days, but four of us stayed 31 days on the mountain! We had rationed our food and were extremely hungry by the end of the expedition. I had started the climb weighing 200 pounds (solid muscle mass) and when I came off the mountain I weighed 150 pounds. This was 50 pounds of muscle loss. I had lost one-fourth of my weight!

As arranged, a bush pilot picked us up at the drop-off point. As soon as we all got off the mountain, we made our way back to Talkeetna. There we took advantage of a shower and shave and then gorged ourselves at a local restaurant for the next few hours. Food never tasted so good!

In terms of donations collected for this event, every cent collected was sent to the organization A Child Is Missing. Expenses accrued for the climb itself were then expensed back to ACIM, so that every penny was accounted for. After the expedition we sold or auctioned all the equipment that I had purchased from the money I raised and sent the proceeds to ACIM. Neither I nor any of my team kept any equipment that had been paid for by funds raised for or equipment donated to ACIM. All the money went back to ACIM. This was also a very successful promotional event for ACIM, designed to help raise attention and awareness about the missing children problem in America.

On an ironic but sad note, my friend and co-climber, Eric Oakes died sometime later after falling from a tree he was pruning. We miss him greatly.

Chapter Twenty

— KEEP SWINGING FOR THE FENCE! —

Trust in the Lord with all your heart and do not lean on your own understanding. In all your ways acknowledge Him, and He will make your paths straight.
Proverbs 3:5-6 NASB

WELL, BY EARLY 2003, the Buckner Building had become more of a liability than an asset. I owed the city of Whittier back taxes and I was beginning to feel the pressure of possible foreclosure. But I had been working hard to find a buyer for the property and it looked like I had one. The Anchorage Daily News reported on February 26, 2003:

> *A New York-based company has agreed to pay $21.5 million for a derelict World War II Army building in rain-slogged Whittier with the goal of turning it into a five-star destination resort. Laidlaw Global Corp., an international finance and investment firm, announced the Buckner Building acquisition this week, noting that shareholders and the Securities and Exchange Commission must still approve the deal.*
>
> *If the deal pans out as planned, Laidlaw executives say they will seek investors and bank financing to pump*

$70 million into the fortress-like building and turn it into a 476-room Ramada Plaza Hotel.[58]

But, unfortunately, it was not to be. Laidlaw Global Corporation was unable to follow through with their funding even though we provided them with substantial appraisals for the valuation of the property. We had contracted Jack Bernholz, a world-renowned appraiser to conduct this 500-page appraisal. In spite of this and other defeats, I could not allow myself to become discouraged or quit. I had to keep swinging for the fence!

I believe that unless one is motivated by Christ, after one has succeeded there is only emptiness and an insatiable drive for the next big win. While the push for the goal is often exhilarating, attaining the goal is usually followed by a let-down. The same holds true for failures. If my purpose is not grounded in something far loftier than a multi-million-dollar deal, then its failure *or* success will leave me unfulfilled, unsatisfied and only wanting more.

I've passed on many seemingly great opportunities, not putting hope in success, but in Christ and what He wants to do in my life. What we *become* is far more important than what we accomplish. A foolish child only wants candy. A loving parent will give that child food that will nourish. I've discovered that if I follow Christ in all that I attempt and do, He doesn't just give me candy, but gives me what will nourish me. He enables me to grow and enjoy life—even in my failures.

Trust in the Lord with all your heart and do not lean on your own understanding. In all your ways

---

[58] Paula Dobbyn, "Whittier's Nuisance Hulk May Become Plush Resort," Achorage Daily News, February 26, 2003, page A1.

acknowledge Him, and He will make your paths straight. (Proverbs 3:5-6 NASB)

All my life, I've been an ardent student of anyone who has faced tragedy and overwhelming circumstances, but has arisen above those to succeed. When I say "succeed" I don't necessarily mean that in the way we would most like to experience it. Money and power are such illusive and fickle measures of success. There are other more noble and lofty measures of success.

We can learn so much by studying others. Take from what others have done and use what's essential. Then discard the superfluous.

Some of my heroes include: Abraham Lincoln, Dr. Martin Luther King, Dale Carnegie, President Harry Truman, President Theodore Roosevelt, President John F. Kennedy, President Ronald Reagan, Audie Murphey, Napoleon Hill, Howard Hughes, Henry Ford, Preston Tucker, Albert Einstein, as well as Jesus Christ, the Apostle Paul, King David, and Moses. I can't get enough of reading about their struggles, failures, and victories.

I've been involved in many if not most of George's enterprises to some degree or another.

George already had a savvy business mind at 15. No one has ever signed George's paycheck. His tenacity and self-confidence and never-say-never attitude are unbeatable.

He's the kind of guy you knock down, back into the corner of the ring and then he'll get up and come out on top. He's resilient. No matter how difficult a situation, he'll make lemonade out of lemons. He's full of confidence, belief, and faith.

George is a very pious man. He is gregarious, exhibits unbelievable faith, and his competence is off the charts.

I knew George before and after he came to Christ. I know him better than I know my own brother.

George has a daunting faith. His faith in Christ is undeniably genuine. He walks, talks and breathes his faith.

– Patrick McCourt, Entrepreneur and friend

Meanwhile, back to my story: Around this time I met a beautiful young woman named Nikki at the Powerhouse Gym in Anchorage. We began dating and eventually became engaged. In many ways we were an odd couple. She was the queen of tree huggers and liberal politics, while I stood on the other side of those issues. She brought her beloved dog, Gypsy into our relationship. Although I love animals, I don't have time to spend with them because of my traveling lifestyle. Nikki was very kind, loving and enjoyable to be with. She had gone to church with me a few times and seemed amenable to the Christian faith.

I will interject here that Nikki had been subjected to a horrible, inhuman ordeal some years before I met her. At the time, she was waitressing at a resort in Alaska. On her way to her car after work one evening, a man snuck up behind her, knocked her out, threw her in the trunk and drove off with her. When she awoke, she found herself in a remote cabin with her abductor.

This demented man force-fed her food laced with drugs. He tortured her and subjected her to unthinkable brutalities and then stuffed her in a freezer. When he returned sometime later, he presumed she was dead, so he took her outside and buried her. In her shallow grave, Nikki regained consciousness and clawed her way out. Then she staggered through the forest to a highway and flagged down help. To my knowledge, her abductor was never found.

Obviously, this bestial treatment and the horrors she experienced impacted her profoundly. As a result I felt a tremendous responsibility to protect her and love her. I wanted to somehow express to her the extreme opposite of what she had endured in her kidnapping.

Nikki loved my daughter Ashley and they both got along very well. One time the three of us went camping on the Kenai River in Alaska. There were signs posted all along the river and in the camp sites "Beware of Bears" and yet, here we were sleeping in a tent! I wasn't particularly afraid of the bears, but maintained a cautious and vigilant attitude toward them. So at night, I slept with my loaded shotgun across my chest. During the night, we heard what surely was a grizzly sniffing about our campsite, but no confrontation occurred.

Sleeping in a tent in the wilderness put Nikki in her element. She often went off into the back country of Alaska for a whole weekend by herself. I hated roughing it in the wilderness! My idea of camping is something much more posh with all the comforts of civilization around me.

Meanwhile, in 2003 I received a call from my good friend Patrick McCourt in Lake Stevens, Washington. Patrick made most of his fortune in property development. He owned more than 10,000 pieces of property throughout the US. He and his team serviced these properties by means of Patrick's two jets and a helicopter. One of Patrick's prime properties was the Streamline Tower of exclusive condominiums on Las Vegas Boulevard in Las Vegas.

He invited me to join him in building and owning a state-of-the-art fitness facility in Lake Stevens, WA. He asked me to design and build this facility to my specifications. Still in Anchorage, I began putting plans together to acquire the World Gym franchise. Having owned the Body Tech Gym and having built the Power House Gym—an award winning gym—I had the background and experience to plan and implement this project.

Patrick owned a prime piece of property in Lake Stevens, Washington, where we would build the first of many fitness facilities around the state. The property was ideally located in a highly dense upper- and middle-class residential area next to the location of a Target store that was currently under construction. Patrick held thousands of properties and had built hundreds of high-end homes in Lake Stevens. Lake Stevens is an exclusive area surrounding the lake by that name just east of Everett, WA.

For this location, I designed a spectacular multi-million-dollar fitness center. But before we began construction on it, the Everett Transit Authority offered Patrick millions of dollars for the property. He couldn't turn down their offer, because it meant making a lot of money without doing anything to the property.

This sudden shift in focus was a big deal to me, even though I was working on other projects simultaneously. I was still trying to find a buyer for the Buckner Building and was looking for a way to revive Alaska Glacier Beverages. I was also occupied collecting from people who owed me money from previous ventures. But I had focused my energies in this property in Lake Stevens, and now the rug had been pulled out from underneath me.

I had spent several hundred thousand dollars in franchise fees and design development. I was now at a loss for producing cash-flow and this put a financial hardship on Nikki and me. This financial pressure weighed heavily on our relationship. The stresses associated with the development of the gym (or lack thereof) coupled with our differences drove a wedge between us and we separated.

The combination of our derailed project and my breakup with Nikki really discouraged me. I had loved her very much and had hoped to spend the rest of my life with her. But I didn't have time to feel sorry for myself and I couldn't allow myself to become so discouraged that it would undermine my future. So, I disciplined myself to move on to the next project.

Chapter Twenty-One
— JOYS AND SADNESS, THE STUFF OF LIFE —

When everything seems to be going against you, remember that the airplane takes off against the wind, not with it.
Henry Ford

AFTER LOSING THAT first property in Lake Stevens, Patrick informed me that he held another property in Monroe, Washington, on which we could build the gym. We conducted feasibility studies and confirmed that the Monroe location would do well also. I had the belief and confidence that we'd do this together. My friendship with Patrick went back many years to our high school days. We thoroughly enjoyed working together. Within several months we built the fitness center in Monroe according to my design. It was a world-class facility!

> After my uncle George came to Christ, he made a complete turnaround.
>
> He's a different man than he was when he owned the night clubs. When he found the Lord, all that junk was eliminated and he brought the family with him.
>
> Since then he focuses on fitness clubs instead of nightclubs.
>
> – Austin LaMoureaux, nephew

However, this time All-Star Fitness heard about our new gym. Once again, Patrick received an offer he couldn't refuse and sold our brand new gym to All-Star Fitness. So here we were again, without a gym. This was extremely frustrating to me, although I could not fault Patrick with the business decisions he was making. Both my business and private life seemed somewhat in disarray.

At this point, Patrick introduced me to a businessman named Greg Ellis. Patrick explained that Greg wanted to buy the Buckner Building from me in Whittier, Alaska. As partial payment to me for the Buckner Building, Greg offered to build me my World Gym. His offer prompted the beginning of a 50/50 partnership together with a plan to eventually build 20 World Gyms in Washington State.

Greg and I located another property in Everett and built our first World Gym there. We pre-sold 3,000 memberships and the gym showed every indication of becoming a huge success. We won the "Best Designed Gym in the World Award" from World Gym International at their annual conference in Ohio. World Gym President Mike Uretz and Lou Ferrigno (the Incredible Hulk) presented me with the trophy.

At that ceremony, I also met Arnold Schwarzenegger and became acquainted with him. Arnold and I continued communication with each other for some time afterward. In 2008, I invited him and Sylvester Stallone to climb Mt. Everest with me. Stallone almost agreed to go with me. And when Schwarzenegger was campaigning for governor of California, he asked me to come down for a portion of his campaign and to attend his inauguration. Due to my commitments, I had to pass on his invitation even though I would like to have attended. I wished him well nonetheless.

But during construction of our World Gym in Everett, we ran into a huge snag. We had the front half of the facility built and nearly ready to open, but we had not yet finished the back half of the building. Greg became enamored with a gold mine in

Alaska and other investment projects. These were very enticing opportunities, so I can understand why he was drawn to them.

After further delays, we finally finished the World Gym facility, but we had lost the better portion of the 3,000 memberships that we had pre-sold. I knew it would take a lot more effort to resell those memberships and Greg and I had a disagreement on how it should be done. So he offered to buy me out and I agreed to let him do so.

Greg did buy the Buckner Building from me, but through a series of financial difficulties that Greg had the Buckner Building came back to me again. When all was said and done, I owned that building for 14 years before relinquishing it to the town of Whittier.

Also, while we were building the World Gym, I created yet another business called Reborn Nutrition whereby I manufactured the product Reborn Gold™. This is a patented, highly proprietary formula with the tagline: "The world's finest, all-natural supplements." I contracted a high-end pharmaceutical laboratory and manufacturing facility to produce Reborn Gold™ to my specifications.

Reborn Gold™ contains every beneficial vitamin, mineral, nutrient, amino acid, and super powerful antioxidant known to man. These include a highly refined proprietary complex of non-metallic silver, aloe vera, green tea and sea-vegetables in a perfectly blended, highly absorbable liquid formula. Reborn Gold™ is all natural, with no artificial colors, no artificial flavors and is ephedra-free. The product contains no thermogenics or stimulants. I trademarked and patented these nutritional products and began manufacturing them on the side. The label goes on to explain:

Reborn Gold™ is made with the world's finest all-natural ingredients, having a substantially high standard of quality and refinement in the nutritional supplement industry. Furthermore, Reborn Gold™ has been designed and manufactured to incorporate an all-natural and very simplistic way to consume daily, every beneficial vitamin, mineral, nutrient, amino acid and super powerful anti-oxidants known to man including a highly propriety complex of non-metallic silver, aloe vera and sea-vegetables in a perfectly balanced formula to help maintain your body's maximum health and allow you the peace of mind in knowing that you have made every effort to give your body the best nutritional supplements in the world today!

I test-marketed Reborn Gold™, produced thousands of bottles of the product, and experienced a huge demand for it. My plan was to initially market the product through my gyms, then take the product national and worldwide. I own Reborn Gold™ free and clear with no other partners.

During these few years in Washington, I was living in a 12,000 square foot home in Mill Creek, driving a black Corvette and enjoyed an incredible lifestyle. I was also focused on building my faith in Christ and finding a godly wife.

For a couple years I attended the Church of Christ in Seattle. This church was really hard-core in the Scriptures. This was an affiliate of the Church of Christ in Alaska. They studied the Bible like a college class. I learned a lot, but the manner in which we studied the Bible seemed impersonal—more knowledge-based than relationship-based.

While attending the Church of Christ in Seattle, I met and began dating Joanne. I had met Joanne at Toshi's Teriyaki Restaurant in Everett, Washington. Toshi's serves great food and it became my evening meal standby for nearly a year-and-a-half. I had never been to the restaurant for lunch, but because I had raved about Toshi's, some of my associates wanted to go there for lunch one day and that's when I saw Joanne. She was a beautiful Korean woman and a daughter of the owner. After my first encounter, I continued to dine there for lunch whenever I could. Each time I saw her, her eyes met mine and sparkled.

However, one day I saw Joanne at Toshi's and said "Hi." But she made a blank face—no sparkly eyes—and I thought, "Well, I guess she must not like me after all." No sooner had I framed that thought in my mind, when the real Joanne walked out of the kitchen and the two women stood side by side. They were identical twins! We all had a good laugh about that. And Joanne's eyes still sparkled for me.

In order to initiate a relationship with Joanne, I decided to send her a single rose that I had a local florist prepare for her. I included my phone number and email on an accompanying card. She responded by emailing me and thanking me for the rose. Apparently, her folks found out about the rose and did not take the gesture kindly, but at the time I didn't know why. I wanted to meet her parents and ask permission to date their daughter, but Joanne barred me from doing so.

Joanne and I dated for about two-and-a-half years and I took her to church with me every Sunday. When I was there, she was there. The church made it known that they would prefer it if Joanne would attend some of the special women's functions. I encouraged Joanne to go to some of these meetings, but

Joanne was an introvert and had insecurities about being with others. The pressure that the church placed on her made her uncomfortable, so we changed churches and began attending Cedar Park Assembly of God Church.

During our engagement we took part in the church's pre-marital counseling and later were married at Cedar Park Assembly of God Church. I really wanted to do things right this time around and that included obtaining her parent's approval, but Joanne was adamant about me not going to her parents. So we married without their consent.

Consequently, her parents didn't know we were married until after our honeymoon. They were extremely unhappy with our union and let Joanne know of their disapproval.

While all of this was going on, my step-sister Janene became very ill and was hospitalized. She was a diabetic and had developed necrosis in one of her feet. This resulted in the amputation of her leg just below the knee. Joanne and I visited Janene many times in the hospital and later when she was moved to an extended care facility.

But shortly after her move to the extended care facility, Janene passed away in her sleep. Her death was to be the pebble that started an avalanche of family tragedies. Only a month later, my nephew, Christopher died in a motorcycle crash. Christopher was my step-brother, Nick's son. Christopher was a fun-loving, handsome young man in his early 20s who had helped me in the construction of the World Gym in Everett.

Christopher had been at a picnic where his boss was showing off his new sport-bike. He offered to let Christopher ride it and Christopher gladly accepted. Only minutes later,

Christopher took a corner too wide and slammed into a tree, killing him instantly.

Some months after Joanne and I married, my work in Lake Stevens came to an end when my partner Greg bought me out. There was nothing else keeping us in Washington, so Joanne and I decided to move back to Anchorage. Our plan was to build a large, new home on Hillside and settle into married life. But while our home was being built we would be living with my parents. Knowing Joanne as I did, I didn't think she would do well under the same roof with my family, so I suggested that she stay with her parents in Seattle until our home was built, but she insisted on going with me.

In Anchorage, I did all I could to help Joanne ease into living in Alaska. Joanne had initially aspired to become a nurse and had attended college in Seattle pursuing medicine. However, during her course of study she realized that nursing and the duties it entailed would pose significant challenges to her due to her tiny frame. Therefore, she changed her focus to becoming a hairdresser and began studying at the Gene Juarez School of Hair in Seattle. When we moved to Alaska, Joanne looked forward to opening her own salon. I had plans to build her a masterpiece salon and help her realize her dream.

So, Joanne and I moved to Anchorage, but in Alaska a number of factors began to weigh down on Joanne. While we were having our home built, we lived with my parents. In spite of my efforts to prepare her for this transition, it came as a shock to her after having the run of a 12,000 square-foot-home in Seattle.

Also, Joanne knew no one in Alaska and it was winter with its long, cold darkness. She suffered under the climate and lack of daylight. For the four years we had been together,

I had tutored her in English. Although she spoke English well, Joanne had grown up in Korea and still clung to her culture and language tightly. And when I asked her to teach me Korean, she refused.

A major part of her cultural upbringing was the prominent role of her relationship with her parents. Before long, I discovered that her parents were systematically driving a wedge between us. In fact, her dad had given her an ultimatum: "Either divorce your husband and come back to your family, or we will disown you!"

Being disowned by her parents was unthinkable to Joanne. To be estranged from her parents like that would have meant near death to her. I tried everything I could to help Joanne work through this. I offered to let her return to her parents for a short time to try to work things out. But she became increasingly overwhelmed with her fear of losing her family and felt compelled to return.

At the same time our fledgling marriage was falling apart, tragedy struck my family again. My brother Bart had struggled with alcohol and drugs for many years. Both Max and I had repeatedly intervened and had taken care of him. We had tried everything we could do to free him from the clutches of those life-destroying chemicals. As early as

---

When my dad Bart died, my uncle George took my brother and me in and took care of us. We lived with him in his big home for about a year and were with him when he was diagnosed with cancer.

I've always loved and looked up to George. When I was a kid, he and my dad took us to Karate lessons. They helped make us strong to become the men we are today.

My uncle George is a great man! He loves his family and would do anything for us. We still get together often and share a lot of laughs together.

– Austin LaMoureaux, nephew

junior high, I remember taking Bart to the hospital to have his stomach pumped due to an overdose.

In recent years, doctors had prescribed oxycodone as a pain killer following Bart's shoulder surgery. Bart was now addicted to that drug and the doctors kept refilling the prescription. I confronted one of Bart's doctors about it and he pretended he didn't know my brother. And later, unbeknownst to me, Bart supplemented oxycodone with heroin.

Oxycodone is nasty stuff that builds up cumulatively in one's system. The crossover going from oxycodone to heroin led to what's known in medical terms as a toxic cascade that took his life. Bart had been very good at hiding his addictions and we didn't know about the heroin until after his death. I was with Joanne when the call came from my brother Max informing us that Bart had died.

Bart's death was one of the most painful events I've ever experienced. I loved my brother deeply—we all did. When Bart died, a part of me died with him. I've never really contemplated suicide, but when he died, I wanted to die with him. His death felt like the greatest failure of my life. Like Jonah in the Bible, I was overcome with my own grief and asked God to take my life and relieve my agony.[59]

Two weeks after Bart's death and a week after the funeral, Joanne left me to return to her parents. I still clung to the hope

---

59 See the account of Jonah in the Old Testament. He was a prophet in Israel, whom God sent to preach a message of repentance to the Assyrians, Israel's arch-enemy at the time. Jonah tried to evade God's mission, but eventually went grudgingly. He preached God's message in the great Assyrian city of Nineveh and the people repented and turned to God. Ironically, this amazing display of God's mercy on Israel's enemies sent Jonah reeling into depression and he asked God to kill him. "I knew that you are a gracious and compassionate God, slow to anger and abounding in love, a God who relents from sending calamity. Now, Lord, take away my life, for it is better for me to die than to live." (Jonah 4:2-3 NIV)

that her parents would renege on their threat and that she would return to me. Joanne and I had lived in our new home for just a short time before she left. I sent her off with her dream car, a brand new $40,000 Subaru Impreza STI. I wanted to take care of her, bless her and desperately hoped she would return. Joanne and I had spent less than six months together in Alaska.

I still have regrets about moving her to Alaska, but her parent's pull was too strong. And it wasn't long before I heard that Joanne was in another relationship. I truly loved Joanne and thought she was the one. We had been together a little over four years.

Since returning to Anchorage, I had been focused on Joanne and building our new home. We had been living off the payments that Greg was making to me from his buyout of my half of the World Gym. I had shipped crates of Reborn Nutrition product up to Alaska, intending to re-launch that business there, but the shipment froze and I lost thousands of dollars' worth of product. So, I put that project on hold for the time being.

Losing all my Reborn Nutrition product was the least of my worries at the time. My grief over the loss of Joanne, Bart, Christopher, Janene, and my business partner overwhelmed me and preoccupied my thoughts.

Chapter Twenty-Two

— THE ODDS ARE GOOD, BUT THE GOODS ARE ODD —

If you learn from defeat, you haven't really lost.
Zig Ziglar

GRIEF IS CUMULATIVE and the combination of having two properties sold out from underneath me, a failed business partnership, my step-sister's death, my nephew's death, my brother's death and now Joanne leaving me all mounted to heap a very heavy weight on my heart.

In the past, I had lost billions of dollars through businesses that were stolen from me or that were not fully funded. Under those circumstances, I had always come back to the family and said, "We just have so much to be grateful for, because all our family is still alive." But now we were losing family members. Without Christ in my life, I don't know how I could have continued to function.

One of the ways Christ helps me cope in life is to maintain a positive attitude, knowing that He is in control, and to keep getting up when I fall. I've found that it helps to stay meaningfully employed and to create something to look forward to.

So, during those months, my brother Max and I created the Mobile Billboard Company. I designed a state-of-the-art truck that served as a traveling billboard. We special-ordered such a truck and used it to advertise Max's insurance company. We intended to order more of these trucks, but the manufacturer became greedy and elevated the price beyond what was reasonable. So we dropped the idea.

I was still reeling from my divorce with Joanne. I truly thought we would spend the rest of our lives together—that was my heart's desire. But now I was single again, and not by my own choosing.

In my loneliness and depression, my nephews urged me to go online and find a wife. I was not at all inclined to meet or date women in this way as it seemed insincere and superficial. In spite of my protests, my nephews kept encouraging me to consider alternate means of meeting women. Although I was hesitant to try the online route, my nephews and other family members weren't so shy on my behalf and entered my personal information on several dating sites.

Not long thereafter I began receiving emails from women who wanted to meet me—over 1200 emails when all was said and done! One of the websites from which I received the most activity originated in Russia and the Ukraine. Many of the pictures of the women that showed up in my inbox looked like supermodels!

Keep in mind, the women in Alaska are few and far between. So, as crazy as it sounds, I flew to the Ukraine for two weeks and actually met approximately 300 of the women I had been corresponding with along with many of their families. The reader may wonder (as I did at first), "How could we possibly orchestrate so many meetings?" But in fact, this Ukrainian dating site had designed a sophisticated process in order to facilitate these initial encounters.

The majority of the women with whom I had corresponded attended an event scheduled for the purpose of introducing prospective daters to one another. Out of those women at this event, I narrowed the field to a dozen, and then to a few, and eventually down to one.

The event was all very well organized and my hosts made me feel like a movie star celebrity! Many of these women were incredibly beautiful. The Ukraine was the breadbasket of the Soviet Union, so apparently they all ate well, got plenty of exercise (the majority of them didn't own cars and walked everywhere) and the result was that they all had great figures!

The event's schedule provided the opportunity for me to date several women. Specifically, I wanted to find out if a woman was a follower of Christ. Also, if I discovered that a woman smoked, drank, or did drugs I did not pursue her. I also wanted to meet a woman's parents.

This whole process really opened my eyes to the greater picture of what was going on here. This foreign matchmaking service is really big business. These sites and services are set up all over the world through which a man can find "an import wife." The women who respond are prescreened to ensure they are not prostitutes. Particularly in the Ukraine, there is a shortage of men and the men there aren't accustomed to treating women well. Consequently, a Ukrainian woman is grateful to find a man to love and care for her. I suspect also that the Ukrainian women find mystique and the prospect for a better future in marrying an American.

The reader might wonder why I would go all the way to the Ukraine for a girlfriend rather than look right here in Anchorage. Let me explain the situation. There is a shortage of women in Alaska. Most of the eligible women in Alaska are young and often leave home when they graduate from high school. The majority of these younger women go to college somewhere else. They want to see the world and usually don't come back. Consequently, there are very few eligible women in Alaska.

In order to demonstrate that I'm not making this up, in 1989, Oprah Winfrey invited "20 bachelors from Alaska—a state known for its abundance of single men—to Chicago to find the women of their dreams. This became one of the highest-rated *Oprah Show* episodes of all time!"[60]

Although undocumented, I have heard that depending where you live in Alaska there are anywhere from eight to 35 men for every eligible woman. The women who are available and looking for a husband quip, "The odds are good, but the goods are odd." Alaska men reply back to that statement warning, "You better be nice to me, or I'll send you back to the States where you'll be ugly again." Most of this is friendly banter; however, there is definitely some truth to both statements.

Regarding the women in the Ukraine, I did have some cause for concern. There are rumors of Ukrainian women marrying an American, staying with him for a couple years and then divorcing him, taking half his fortune with her and returning to her family in the Ukraine. I would have had to be totally convinced of the authenticity of a woman's love. I thought of the old Memorex cassette tape commercial, "Is it live, or is it Memorex?" In other words, "Is your love genuine, or is it a ploy for a Green Card?" I needed a lot of wisdom in this matter.

---

60   http://www.oprah.com/own-where-are-they-now/Alaskan-Bachelors-Looking-for-Brides-Video.

Had I actually agreed to bring a woman home from the Ukraine, I would have done so by means of a Fiancée Visa. In this way, we could have spent time together, getting to know one another before committing to marriage. At the time, however, I had no peace about moving forward and in view of what soon occurred it was probably for the best.

When I returned to Anchorage from the Ukraine, I needed to generate some quick cash. A friend of mine suggested I get into real estate. He was making a lot of money, so I became a real estate agent with Prudential and did extremely well. I really threw myself into real estate, advertising properties in the Real Estate Magazine and running as many as 600 television commercials per month. As a result, I was a top realtor at Prudential, selling some of the most exclusive properties in Alaska. The last two houses I sold went for $1.6 million and $1.4 million respectively, earning me handsome commissions.

As I look back on my life at that time, I needed a new focus and a clear direction for my life. Sometimes we need a sobering, monumental event to redirect our attention toward that which is truly significant in life—something that enables us to see with crystal clarity what is essential and eternal. As if my other recent experiences weren't enough, I was about to be jolted by such an event.

Chapter Twenty-Three
— CANCER —

How do you nurture a positive attitude when all the statistics say you're a dead man? You go to work.
Patrick Swayze

In September 2007, I caught what I thought was a cold that I just couldn't shake. I went to a first-care clinic and got a prescription for a round of antibiotics. We were pretty sure I had an infection and assumed the lump on my neck was due to my lymph nodes fighting the infection. Years ago I had experienced a similar swelling and assumed that this too would be solved with antibiotics.

The first course of antibiotics didn't kick the infection, so I went back to the clinic and asked for another round of antibiotics. The clinic pushed back and wanted to send me to a specialist, but I talked them into giving me more antibiotics, so I took a second round—still to no avail.

At this point, the clinic referred me to an ear, nose and throat specialist, Dr. Rosane in Anchorage. He took a biopsy of the lump on my neck. When the results came in, he called me for an appointment. I could tell he was hedging, but I finally got it out of him that I had squamous-cell carcinoma of the head and neck. The cancer had already progressed to stage four.

In response to this news, I hurled myself into intense research and unearthed thousands of pages of information on squamous-cell carcinoma. What I learned is that if I endured the surgeries, chemotherapy and radiation treatments, I had a 50 percent chance of surviving more than five years. If I refused all three of those treatments, the prognosis was that I would live half that or less.

As is often the case with life-threatening and chronic illnesses, they precipitate other catastrophes that cascade into an avalanche of woes. In my case, when I notified my health insurance carrier of my cancer and the pending surgery, my insurance company simply dropped me. Even though I had been paying my premiums, suddenly I was without medical insurance. I probably could have sued the insurance company, but it would have been a tough battle and I had no energy to pursue it. No doubt they took that gamble when they cancelled my coverage.

Even though the nature of the challenges I had faced up to now had been huge including failed business ventures, broken relationships, losing my wife, and the untimely deaths of my step-sister, my nephew, and worse my brother Bart—all within the last three months. But with those challenges, regardless of the outcome, my life would still go on. I could always get up again, rally my talents and resources and go on to the next thing. Now I was fighting for my life. As a friend of mine says, "You never see life more clearly than when you're staring death in the face."[61]

I put real estate and every other business venture on hold for the time being. I needed to focus solely and intently on beating cancer. Dr. Rosane scheduled my first surgery for October. In that surgery he removed the lymph gland from my neck.

---

[61] Rob Fischer

Following the surgery I puked my guts out. The procedure really took a lot out of me leaving me weak and feeling beat up.

Watching my physical strength and stamina diminish like this was really a blow to me. I had always been extremely health-conscious and fit. I was a black-belt martial artist, weight-lifter and runner. I had never abused my body with smoking, drinking or drugs. Physical strength and prowess was at the core of who I was. I had always trained to win. But now, I felt weak and vulnerable.

I refused to take pain meds after this first surgery and after all of my subsequent surgeries. The way I figure it, God gave us pain as an indicator like the "check engine" light on a car. If we mask the pain (i.e., ignore the warning light) we haven't solved the problem. I wanted to maintain full control of my faculties so I knew exactly what my body was experiencing and how I was reacting to the treatments.

Going into the first surgery I was able to maintain a pretty positive outlook. I thought, "I'll go in, get it done, and get out." But when Dr. Rosane informed me that I needed a second surgery I grew more concerned. Following that second surgery in November, my fears mounted.

After the second surgery, the doctor wanted to conduct a horribly invasive surgery on me in which he would cut my jaw open and lay both halves back in order to gain access to my throat. He also wanted to remove half the muscles in my neck and shoulders and take half my jaw. This procedure would have left me unrecognizable even to family members. Frankly, this surgery sounded like a lousy second option to death and

I asked for another opinion. Dr. Rosane recommended a specialist in Seattle.

That December I flew to Seattle and stayed with my friend Patrick McCourt. The specialist at Virginia Mason Hospital recommended three more surgeries to remove parts of my tongue, throat, and more lymph nodes. Rather than experience three separate surgeries, I elected to have all three surgeries performed at once.

After the surgeries, my chance of surviving another five years was only 50 percent. As I lay in the hospital bed with hoses draining fluid out of my neck, I had a significant experience with God. I lay there knowing that my life hung in the balance. In spite of the surgeries, death was still a looming possibility.

In that hospital bed, I recommitted my life to my Lord and asked Him how I might serve Him and bring Him glory before I die. My prayer was not a desperate "foxhole" plea for deliverance: "God, if You'll get me out of this mess, I'll live the rest of my life for You." No, I had already been living for Christ to the best of my ability.

My prayer now was of a different sort. I desperately wanted to leave a legacy for others. I wanted to leave an eternal mark for others—something far more transcendent than a huge business deal. Business, money, stuff—it's all temporal. What really counts are people. People are eternal. I wanted to leave a legacy that would benefit others in some profound way. When I died and stood in God's presence, my deepest desire was to hear Him say, "Well done, good and faithful servant." (Matthew 25:21)

From my hospital bed, my prayer was, "Lord, I don't have much time left. What can I do to leave a legacy for my family, for others and bring glory to You?"

When I prayed that prayer, I sensed God prompting me to finish the Missing Children's Telethon. To do so meant that I needed to come up with something sensational to attract the attention of the community and viewers. I knew that people simply would not watch a Missing Children's Telethon unless it was woven into a sensational and daring film event.

In order to appeal to the secular community and attract viewers by the millions, I needed a monumental event—something incredible with death on the line. By entertaining people to the max, I knew I could hold their attention and then spoon-feed them information about ACIM.

I thought about how to continue what I had started with my climb of Mt. McKinley. But whatever I tackled next needed to be increasingly more perilous and adventuresome. At first, I did not know what I was going to do, but knew that I had to finish this effort to raise attention and awareness about the missing children problem here in America and around the world.

As I pondered how to promote ACIM's cause, the thought came to me, "Climb Mt. Everest." Immediately, my goal became crystal clear: I would climb Mt. Everest and film the expedition with the intent to create a telethon on behalf of ACIM.

Prudence would have argued waiting at least a year until I was fully recovered. But I felt that waiting would have been more self-promoting and less God-glorifying. So, I opted to take the risk of climbing Everest only three months after my fifth cancer surgery. In this way, my success would be testimony to *God's* ability to make this happen rather than my own ability.

> More than anything else, Christ was the one that helped George get through cancer.
> – Terri Lane

Lest there be any misunderstanding, I was not (am not) an avid mountaineer. I had only climbed two mountains in my life: O'Malley Peak and Denali, and both of those climbs were motivated by a desire to raise funds for worthy causes. I didn't even own any mountain climbing gear. Climbing Everest was not some secret passion of mine. Mt. Everest was not on my "bucket list." At that time I had never even heard of the concept of a "bucket list." (The movie, *The Bucket List* came out at Christmas 2007 after I had already decided to climb Everest.) The concept of a "bucket list" had never even entered my mind.

At this point, let me offer two pieces of advice for someone fighting cancer. (Later, I'll offer a third recommendation.) First, if you've never done so, put your faith and trust in Jesus Christ. Putting one's faith in Jesus Christ should not be "our last ditch effort" when facing cancer or some other tragedy. But it's okay if that's what it takes to bring us to Him. He loves us and wants a deep, personal relationship with us and we all need His forgiveness and cleansing from our sin.

> *For God so loved the world that he gave his one and only Son, that whoever believes in him shall not perish but have eternal life. – John 3:16 NIV*

I also want to be very clear about what I mean by "trusting Christ." I am not referring here to some New Age nonsense about tapping into an impersonal spiritual power. I and my family had already tried all that and found it phony and lacking any real substance. What I'm talking about is a real relationship with the real Person of Jesus Christ—our Creator, Sustainer, and Savior. "In him we live and move and have our being." (Acts 17:28 NIV) In Christ we have access to real redemptive, creative, and restorative power.

As I faced cancer, I already knew Christ. I don't know how I would've managed without Him. I had just lost my step-sister, my nephew, and my brother, and then my wife left me—all in the last three months. And now *my life* was slipping through my fingers. All of this amounted to a ton of grief and anguish.

Even though I walk through the valley of the shadow of death, I will fear no evil, for you are with me.
– Psalm 23:4 ESV

Cancer is a life-altering battle with many dark alleys and skirmishes along the way. Fighting cancer impacts every facet of your life: work, finances, family, relationships, goals, routines—it affects everything. Christ was ever present with me, encouraging me and strengthening me through this ordeal.

I can do all things through him [Christ] who strengthens me. – Philippians 4:13 ESV

The second course of action I would recommend for someone facing cancer is to find a transcendent cause or goal—a way to serve others and glorify God in the middle of your battle for life. Stay away from all tendencies to play the victim. Once you go there and take on a victim mentality there is no victory. Victims remain victims of their circumstances and environment. A victim never rises above their situation; they're always caught in it—beaten down and destroyed. A victim will always view himself as victimized—controlled by the situation and without any power to change that.

A victim is consumed with self-pity, which is nothing more than self-focus and arrogance in disguise. Victims give in to despair. That's the downward slope where the victim mentality leads. Such thinking is never restorative or healing in nature, but destructive and disempowering. The way out of the pit of despair and out of self-pity is to serve others.

Find a transcendent cause that will benefit people and then throw yourself into it with abandon. Be consumed with a desire to make life better for others. For me, climbing Mt. Everest on behalf of the children and their families represented by ACIM was the gift God gave me to rise above despair and self-pity.

Chapter Twenty-Four
— NEVER QUIT! —

Right on the other side of perseverance and endurance is God's blessing.
Dr. Kenneth Friendly

AFTER THAT TRIPLE surgery, which comprised my third, fourth and fifth surgeries, I remained in Virginia Mason Hospital in Seattle for three or four days. Then I moved back into my friend Patrick McCourt's home and stayed with him for a couple days.

I think Patrick wanted to help me take my mind off the cancer by offering me a lucrative business proposition. He had one-and-a-half billion dollars in assets that he wanted me to liquidate for him. This is a specialty I have and normally I would've jumped at the opportunity. If successful, this work could've earned me $45 million in commissions.

But I explained to Patrick that I needed to first focus on climbing Mt. Everest on behalf of the children and families represented by ACIM. Then, after the climb, by God's grace I'd help him liquidate his properties. So I got on a plane and headed back to Anchorage. On the flight, my head and neck were wrapped in huge bandages and I was still *very* weak.

It was good to be back home in Alaska with my daughters Ashley and Tamra and near my mom and dad and the rest of the family. They were such an encouragement to me and I gained so much strength from simply caring for Ashley and spending time with my family. I should mention that after my brother Bart died, his two sons had nowhere to live, so I had taken in my two nephews and they continued to live with us as well.

At this time, my thoughts also returned in particular to one beautiful woman whom I had met in the Ukraine. I had considered bringing her back to the US on a Fiancée Visa to date her and see how our relationship might develop. But I hadn't been able to follow through with her due to my cancer. (I later learned that she had married someone else.) Now, battling cancer, finding a wife became substantially less important to me. I just wanted to be with my family and spend the final days of my life with them—especially with my daughters. In one respect, finding a wife was one more thing I didn't have to worry about for now.

This was a *very tough* time for me. My weakened physical condition and cumulative grief sapped my energies and seriously threatened my motivation. At the same time, I felt an overwhelming sense of responsibility to care for my family—and especially for Ashley, Tamra and my two nephews. I was the only one working in the household and bills were mounting. As I mentioned, I had hoped to re-launch Reborn Gold™ in Alaska and had shipped all the product from Washington to Anchorage. But my shipment froze, exploding the bottles and I lost thousands in product, so a re-launch of Reborn Gold™ at the time did not seem viable.

I felt like I was being crushed under a mountain of stress at this time. As I listed the responsibilities that were on my plate, I had to: 1) nurse myself back to health and fitness to be able to climb Mt. Everest in two-and-a-half months; (I sifted through thousands of pages of research on cancer to understand what was happening to me. I also called Carol, my former wife and Alternative Health Care Practitioner and asked her for her medical related recommendations.) 2) Join or create an Everest climbing expedition; 3) raise support for that expedition; 4) figure out how to pay for my cancer surgeries; and, 5) earn enough money to pay for everyday living expenses and care for my daughter and nephews.

Of the above responsibilities, the obvious one to eliminate would have been climbing Mt. Everest. It would have been so easy to cut that massive undertaking from my agenda and experience a measure of relief. And believe me, I considered doing just that. But about this time, a friend of mine from ChangePoint church gave me the Discovery Channel DVD series, *Everest: Beyond the Limit*.

This series utterly captivated me! Deep down inside, before watching *Everest: Beyond the Limit*, I knew I would climb Everest, but I didn't know how. Watching this series inspired me to move ahead and focus on reaching this goal. I contacted the expedition leader featured in the TV series, but his north face climb on Everest had been preempted that year by the Chinese government. The Chinese had taken over the north face route in order to take the Olympic torch to the top of Everest. As a result they cancelled all other scheduled north face climbs that season.

Watching the Everest series prompted me to conduct massive research on the technical aspects of the climb. I read thousands of pages on Everest to understand all that was involved and what it would take for me to summit that mountain.

I continued to call every other Everest guiding service I could find. It was too late in the year and too costly for me to assemble an entire team, although that's what I had wanted to do. I was afraid that if I merely joined an existing team that my chief purpose of raising awareness and funds for America's Missing Children would be lost in the mix of each team member's personal goals for climbing Everest.

Fortunately, I found Peak Freaks and was able to join an existing team on which I became the seventh of nine members. God was most certainly with me in this regard! Obtaining a place on an Everest climbing team is very difficult to begin with, much less three months prior to a climb. Normally, people make plans and train for years in order to climb Everest. So what I accomplished in the amount of time given was truly out of the ordinary!

Also, I was able to recruit all the team members of my team to climb on behalf of ACIM. To be sure, each of the other members had their own personal goals and reasons for climbing. But my team members all joined me in promoting ACIM. The team was to be led by Tim Ripple who had 16 years of experience in the Himalayas at the time with a good safety record. For seven of those years Tim had guided people up Everest, but had yet to summit himself.

With that said, I was taking a leap of faith going with Peak Freaks and Tim Ripple, however, I knew—just as I had on McKinley—that I would stay on Everest until I had summited. Also, I must confess that Peak Freaks knew nothing about my cancer until I arrived in Nepal. Otherwise, they wouldn't have let me climb with them. I had obtained a doctor's statement saying that I was healthy, but that statement neglected to mention that I was battling cancer.

Due to the narrow climbing window on Everest (only a few weeks each spring), the schedule really put me under pressure. Less than three months after my last cancer surgery, I would be on my way to Kathmandu, Nepal, beginning the arduous trek into the Himalayas toward Everest!

With this lofty goal before me, I needed to obtain the funding, regain my health and strength to prepare for the rigors of a Mt. Everest climb. But I found that I was tired all the time—just wiped out. Because I had invented Reborn Gold™, I already had a good working knowledge of what I needed to do to regain my health and strength. (I'm convinced that had I been taking Reborn Gold™ this whole time, it most likely would have prevented my cancer based on what I had learned about what it would take to heal me, but my supply had been destroyed.) With this knowledge, I went on the search for resources in Anchorage and found Steve Planté, owner and proprietor of the Organic Oasis Restaurant and Juice Bar.

Steve's personal story of overcoming lead poisoning through diet is inspiring. Steve follows the ancient advice of Hippocrates in this regard, "Let food be thy medicine and medicine be thy food." After Steve experienced his own journey back to health, he "decided to help others learn how to use food as a weapon against disease." [62]

---

62 http://organicoasis.com/about.

In order to regain my health and strength, I knew I needed to eat strategically. But I also knew that on top of everything else I was doing, I didn't have time to prepare my meals like I needed to. So I met with Steve that first time for a couple of hours and simply shared my story. I told Steve about my cancer and how I was going to climb Mount Everest in March to benefit ACIM. Steve committed then and there to support me through the process—that is, he offered to feed me and send me to Mt. Everest with what I needed to regain my health and strength. Steve's generosity and care for me was a tremendous boost to my morale and my health!

Additionally, the healthy diet he put me on did wonders for me in my recovery. Steve told me, "Conventional medicine is addressing your symptoms, but not the core issue. You need to build up your immune system so it can effectively fight the cancer and you need your strength back. We can do that with nutrition." This is the third piece of advice I would offer someone facing cancer: get onto a nutritionally healthy, high alkaline diet and stay with it.

I began eating all my meals at the Organic Oasis. Steve helped me establish a nutritious diet that put me on the road to recovery. With Steve's guidance, I could have prepared this diet for myself at home, but I didn't have time to work and train intensely and do everything else and still prepare my meals. I could not have done this without Steve's help. Financially, physically and mentally I could not prepare my meals without Steve. Steve's help was a miracle in itself!

Steve and his staff put me on a regimen to show up twice a day for a *high-protein platter* that they had designed for body-builders and athletes. The high-protein platter consisted of two grilled organic chicken breasts, lamb, or other organic protein, a cup of organic steamed vegetables, and a cup of organic quinoa. While that was being prepared, they whipped up a power smoothie for me made with organic apples, organic bananas, organic strawberries, flax oil, hemp oil, bee pollen, trace minerals and lecithin. I would down this smoothie as my appetizer.

On this strict diet I craved dessert, but they only allowed me one scoop of organic ice cream about once every other week. They explained that cancer loves and thrives on sugar and that one of the keys in any cancer therapy is to eliminate sugar.

In addition to the above, they set out little cups of supplement capsules for me to take containing antioxidants—specifically resveratrol (extracted from grape leaves) and astaxanthin. They also gave me what Steve called "The five pillar stones of the immune system." Those five are: beta carotene (marine source), vitamin C (buffered), vitamin E, zinc, and selenium. (Steve has since added vitamin D3 as the sixth pillar.)

Steve explained a well-known fact that cancer cannot live in a body with a PH above 7.4, so raising my PH levels was one of the goals. They also added fish oil to my meals because

> *Several years ago, George came into the Organic Oasis and shared his story. I don't know how he found us. But we've been helping people find their nutritional baseline since 1987.*
>
> *George told me about his cancer and that he planned to climb Mt. Everest in three months.*
>
> *We told him that if he wanted magnanimous changes in his body, he'd have to make magnanimous changes in his diet and lifestyle.*
>
> *He had an incredible can-do attitude. I told him, "Your life has been a party up until now, but you're going to have to change all that."*
>
> – Steve Planté, Owner, Organic Oasis Restaurant and Juice Bar

the high protein platter is low on fat and without it my body would have craved the not-so-good fats. Finally, they gave me a multivitamin and a B-complex. Steve told me the B-complex was necessary because, "You're such a hard charger and lead a high-stress life!"

As I began eating in accord with a cancer therapy that Steve put together for me, I began feeling better and gaining my strength and energy back. (I was pleased to note that Steve's recipe of supplements corresponded with those that I had assembled in Reborn Gold™.) But after my bout with cancer and the five surgeries, I also needed to rebuild muscular strength and stamina in preparation for the rigors of Everest.

If I was going to climb Everest in March, I needed to be in top physical condition and I sure didn't feel like that after my surgeries. I needed focused strength and aerobic training. I also knew that physical exercise releases endorphins and helps improve one's mental outlook. So exercise would have a dual positive effect on me. My cardiovascular training consisted of doing laps up and down the stairs in my home. Before leaving for Everest, I was up to 500 flights of stairs per day. I had purchased a hand-held counter in order to keep track of my stair laps.

This diet and exercise regimen sounds pretty tame as I write about it now, but it was very tough on me. This was a difficult time. Consequently, I also gave a lot of attention to my mental and spiritual well-being. Since coming to Christ, I have always gained strength through reading and listening to God's Word, the Bible. I also find encouragement and strength by attending church and listening to sermons.

I've listened to many great preachers during my walk with Christ including Adrian Rogers, Alistair Begg, Jerry Prevo, Dan Jarrell, Karl Clauson, Kenneth Friendly and the list goes on and on. I happened to be listening to Dr. Ken Friendly's radio program through Lighthouse Christian Fellowship in Anchorage, when he preached a four-part sermon series called, *Never Quit!* Dr. Friendly preached that message with eloquence, fervor and animation. He is a genius and his message continues to inspire me. I've been exposed to hundreds of courses on motivation and success, but Dr. Friendly summarized those core principles so articulately.

Dr. Friendly's message, *Never Quit!* spoke to me so powerfully and profoundly that I purchased the CD series and listened to it over and over again—perhaps a hundred times! In fact, as I performed my laps on the stairs, I broadcast Dr. Friendly's message through the sound system in my home, booming its energizing truths into my core. (I trust that Ashley and my two nephews also benefited from Dr. Friendly's message as they could not escape hearing it!)

Dr. Friendly's talk gave me strength and inspired me to keep going. His message was as much a part of my preparation as my diet and exercise. In that message he cited such maxims as:

Quitting is the number one killer of dreams and visions.

A person who habitually quits is a loser. You're not a loser because things don't turn out, you're a loser if you quit!

Every godly resolve will be challenged. In fact, all hell will break lose!

Right on the other side of perseverance and endurance is God's blessing.

Hold your course, be a finisher.

Though the righteous fall seven times, they rise again.
– Proverbs 24:16 NIV

I still listen to Dr. Friendly's message from time to time and gain encouragement and strength from it. [63](To read an abbreviated version of Never Quit, please see the Appendix at the back of this book.)

Besides caring for my physical, emotional and spiritual well-being, I committed myself to raising money for the Mt. Everest climb on behalf of A Child Is Missing. Naturally, I had called Sherry Friedlander to pose the idea to her and she was very excited about the fund-raiser and what it could mean for ACIM. So I had her blessing for the climb.

Meanwhile, I began visiting business people and acquaintances in Anchorage to raise support for the climb. Rod Udd at Anchorage Dodge and Chrysler put up the first $1500 and many others pitched in generously. I also continued selling real estate simply to pay my bills. It was difficult working and raising money while trying to get well again.

As the date for my departure to Nepal drew nearer, I still lacked the necessary funds for the Everest expedition, so I took out a home equity loan for $50,000 to make up the difference. Everything was quickly coming together for my departure.

---

63 Dr. Kenneth Friendly, *Never Quit*, Anchorage, Alaska: Lighthouse Christian Fellowship, 2008. (To order Dr. Ken Friendly's four-CD message set, *Never Quit*, go to www.lighthousealaska.org or call 907-272-2252.)

Chapter Twenty-Five
— MT. EVEREST —

Effort only fully releases its reward after a person refuses to quit.

Napoleon Hill

IN LATE MARCH, 2008, I flew to Kathmandu, Nepal and stayed there for a day or two to adjust to the time zone and see the sights. As I arrived, the Nepalese had been celebrating a "color festival," which gave Kathmandu a playful atmosphere.

From Kathmandu, I took a small plane to the Tenzing-Hillary Airport in Lukla about 85 miles east. Two months before my arrival the airport was renamed in honor of Sir Edmund Hillary and Sherpa Tenzing Norgay, the first climbers to summit Mt. Everest.[64]

At 9,200 feet above sea level, the Tenzing-Hillary Airport is considered by many to be the most dangerous airport in the world. The airport's paved runway slopes at a 12 percent grade and stretches only 1,510 long by 66 feet wide.

> Before he left, I whipped up some energy bars for George. I made about 100 bars for him. They looked like hockey pucks. I designed them to be easy to digest, but loaded with calories, both of which were vital for a high altitude climb.
>
> I took George on because he had the secret ingredient: attitude! Most people don't realize how important attitude is to overcoming cancer.
>
> – Steve Planté, Owner, Organic Oasis Restaurant and Juice Bar

---

64   http://en.wikipedia.org/wiki/Tenzing-Hillary_Airport.

The south end of the runway drops off steeply to the valley below and the north end butts up against high terrain.[65] There are no "go-arounds." Landing and taking off from there was a blast and felt like an amusement park ride!

Like many other mountaineers in the Himalayas, we started our long trek to Mt. Everest from Lukla. Trekking to the Everest Basecamp from Lukla is only about 38.6 miles. But the trek requires eight to nine days on the ascent, which enabled us to acclimatize to the altitude and continue our physical training. The route follows a narrow, winding trail through the mountains climbing over 8,000 feet.

The trail connects about eight small villages with names like Monjo, Namche, Khunde, Tenbouche and Pangboche. These villages offered "tea houses" in which we bunked and received tea and a hot meal. The tea house bathrooms consisted of either a hole in the ground or a bucket. The tea houses provided the locals with a great source of income during the climbing season. We found the Nepali people to be incredibly hospitable and kind.

During the climbing season, this trail is crammed with climbers, Sherpas, heavily laden yaks and trekkers. Everything that goes in and out of the villages is transported either on the back of a Sherpa or a yak. I watched one Sherpa plod up the trail with a full-sized washing machine tied to his back. Another carried a table and four chairs as his burden! And watch out for the yaks! They have the right of way. A yak ambled into me one day knocking me off the trail.

---

65 http://en.wikipedia.org/wiki/Tenzing-Hillary_Airport.

From Lukla I had trekked out with one of my team members, Dom Gilbert from Canada. We had arrived a day or two behind the other members of the team, so it took us a few days to catch up with them. On March 30, we joined other members of the team in the village of Dinboche. Tim Ripple, our expedition leader was not there. He was trying to get our permits sorted out due to major restrictions placed on climbers by the Chinese.

One of our team members, Scott Mortensen from the US, sent a communique about my arrival with Dom back to Becky Ripple, Tim's wife and support partner back in Canada:

George and Dominique arrived last night here in Dingboche. Dominique seems to be suffering from the same stomach flu that I dealt with. George is on the other end of the spectrum, full of energy and passion. His mission to Everest is to film our expedition in attempts to raise funds and awareness for America's missing children. He has two great HD cameras in his pack which is wonderful, since mine died in Pangboche.... another victim of altitude sickness I guess. Anyway, in a simple twist of fate George is without a cameraman/writer for his mission...so it seems I've picked up some more work. My current goals are to shoot some Everest footage for an independent film I'm working on, cut together a trailer for Tim and Becky of Peak Freaks, help out George with www.climbforamericaschildren.com, and oh yes, summit Mount Everest.[66]

I had Jerry Neeser, owner of Neeser Construction, to thank for those cameras that Scott mentioned me carrying.

---
66 http://peakfreaks.com/everestnews2008.htm.

Just days prior to my departure from Alaska, I still had no cameras for the expedition. Jerry and I had met at the Organic Oasis. Through Jerry's very generous donation, I was able to purchase top-of-the-line Sony HD digital cameras in order to document the trip.

The thin air made the trek to Base Camp more difficult than one would think. Once on the mountain, we continued to acclimatize by climbing to higher elevations during the day and coming back down to sleep at a lower elevation at night. As our bodies allowed, we climbed higher and began sleeping higher. All of this was mandatory to achieve proper acclimatization and avoid altitude sickness.

Altitude Sickness, often known as Acute Mountain Sickness (AMS) is particularly an important consideration while trekking in Nepal. Altitude Sickness means the effect of altitude on those who ascend too rapidly to elevations above 3000m. The initial symptoms of AMS are as follows:

- *Nausea, vomiting*
- *Loss of appetite*
- *Insomnia / sleeplessness*
- *Persistent headache*
- *Dizziness, light headache, confusion*
- *Disorientation, drunken gait*
- *Weakness, fatigue, lassitude, heavy legs*
- *Slight swelling of hands and face*
- *Breathlessness and breathing irregularly*
- *Reduced urine output*

These symptoms are to be taken very seriously. In case of appearance of any of the above symptoms any further ascent should be reconsidered; otherwise more serious problems may occur which can even cause death sometimes within a few hours. The only cure for the Altitude Sickness is to descend to a lower elevation immediately. Acclimatization by ascending no more than 300 to 500 meters per day above 3000 meters and the proper amount of rest are the best methods for preventions of AMS.[67]

Less oxygen, lower barometric pressure, rapid ascent, dehydration, and hypothermia are all causes of or factors that contribute to altitude sickness. High Altitude Pulmonary Edema (HAPE) and High Altitude Cerebral Edema (HACE) are more serious manifestations of altitude sickness and can end in death unless the victim descends to lower elevations quickly. Knowing these things and following proper procedure were keys to our safety.

Tim kept us on a strict ascent schedule to minimize the possibility of AMS. We had all been thoroughly briefed on other precautions like drinking plenty of water, no alcohol, no tobacco and keeping our calorie intake high. I "drank water like a fish and ate like a pig" the whole time I was on Everest. This strategy is a critical factor in climbing at high altitudes, because the desire to eat and drink falls off sharply. But hydrating and fueling the body is precisely what one must do. I also refused altitude sickness preventative drugs and instead opted to regulate my body with plenty of fluids, as

---

67 http://peakfreaks.com/ams.htm.

Today we would like to introduce the eldest member of our team - George LaMoureaux 51 years old, a realtor from Anchorage, Alaska. George's mission for climbing Everest is for the purpose of filming, "The First National Missing Children's Telethon" going from the "Coldest Mountain on the Planet" (Denali) to the "Tallest Mountain on earth" (Everest) and back to the "Studio Audience: Lost and Found Children, Law Enforcement, Supporters and Celebrities' Endorsements along with Film Shorts for Peak Freaks." Scott Mortensen, Tim Rippel and Sherpa supporters will be behind the camera.

LaMoureaux has been on a seven year mission to help find children and is also fighting another battle. He is a "Recent Cancer Survivor" who has had five Cancer Surgeries in the last three months and is on a "Customized All Natural Nutritional Recovery Program" for pre-climb, during the climb and post climb of Everest

much organic food as possible, and all of the supplements I had brought with me.

Of course there are many other perils on Everest besides the low oxygen altitudes. Add high winds, severe cold, snow storms, crevasses, a moving icefall, avalanches, cornices, equipment failure, sickness or injuries, frostbite, and human error and we've got a recipe for possible disaster. It's no wonder that for every ten successful ascents on Everest there is one death.[68]

In one of the villages on our trek to Base Camp, I received an extremely disturbing call from my family via satellite phone. My little sister Mia had been found dead on a woman's deck under suspicious circumstances. I was shocked and grieved for Mia and my family. All I could think about was getting back to my family and trying to sort out her mysterious death. But my family urged me to finish the climb, reasoning it was what Mia would have wanted. From then on, I dedicated the climb to my sister Mia. I suspect that

---

68   Stewart Green, Death on Mount Everest, www.Climbing.About.com., 2012.

the hardships of the climb helped me work through my grief.

Our team proved to be a colorful blend of adventuresome individuals. In addition to Tim Rippel as expedition leader, we were honored to have Farouq Alzouman, who would become the first Saudi to summit Everest. Sultan Al-Ismaili from Oman hoped to bring his country the same honor. We also had Larry Williams, a school teacher from the US; Scott Mortenson from the US, who would film the event; Dominique Gilbert from Canada; Nabil Lodey, retired British military serving as an international law attorney in France; and Saad Naseer from the US. Saad did not join the team until later as he was vying for a rapid ascent record. (Having just come off another high-altitude climb, Saad was already acclimatized to the altitude.)

from sponsor Steve Plante', owner of the Organic Oasis. LaMoureaux, also received substantial recognition and support from the Lance Armstrong Foundation for inspiring and empowering people affected by cancer.

LaMoureaux summited the 20,320 ft. peak of Denali in 2002, which was filmed by a FOX / Olympics Camera Crew. He and his team went up Denali with only 17 days of food and because of bad weather conditions stayed for 31. Consequently half the team turned around and went home, but LaMoureaux stayed and summited with the other half of the team, which will be a part of the broadcast event. Good luck George! There are many prayers coming your way.

– Becky Ripple, Peak Freaks

Although it's difficult to substantiate a claim such as this, based on the Fox News story featured on October 6, 2014, I may well be the first cancer survivor to have summited Everest.[69]

During the first week of April, all of our team except for Saad had arrived at Base Camp. We were the first team to

---

69   Fox News, "Top of the World: Meet the First Cancer Survivor to Climb Mt. Everest," October 6, 2014, http://insider.foxnews.com/2014/10/06/top-world-sean-swarner-becomes-first-cancer-survivor-climb-mt-everest.

arrive at Base Camp, which was a blessing because it would soon swell into a tent village. Here we focused on safety and rescue training and began preparing for crossing the notorious Khumbu Icefall.

During the many weeks we spent on Everest, we would end up crossing the Khumbu Icefall numerous times and often twice a day. This was necessary for two reasons: first, the South Col route demanded it; and second, the Icefall lay just above Base Camp. In order to acclimatize, we had to climb to higher elevations from Base Camp and then return, necessitating all these treks through the Icefall.

The Khumbu Icefall, with its house-size ice blocks, deep crevasses and tall, sharp seracs has claimed many lives over the years. On one descent, I came all too close to joining that number! And on our first day working our way through the Khumbu Icefall, we encountered the frozen remains of some poor unfortunate climber whose life the Icefall had claimed years before.

Chapter Twenty-Six
— WAITING ON THE CHINESE —

People often say that motivation doesn't last. Well, neither does bathing - that's why we recommend it daily.
Zig Ziglar

FOR MORE THAN a month, our team and other climbers on Everest were subjected to the ever-changing whims of the Chinese government. The Chinese hosted the summer Olympics in 2008 and had decided they wanted to light the Olympic Torch on the summit of Everest and carry it to the games. In order to do so, they shut down the mountain to *all* other climbers while they were on the mountain. In addition, they completely closed the north route for that climbing season. A large contingency of Chinese military were on the north slope of Everest preparing for their ascent. We would be climbing the south slope.

The Chinese had approximately 200 climbers and 200 Sherpas taking over the whole north side of the mountain. Compare those numbers with our team of 14—seven climbers and seven Sherpas. But the Chinese banned *all* other activity on the mountain above Camp 2 through May 10. If the Chinese delayed their summit bid too long we would miss our entire window of opportunity to summit this year rendering our climb a failure.

The Chinese were very serious about keeping everyone off the mountain and told us that their snipers had orders to shoot any climbers who attempted to climb during the shutdown. I spoke with one of the snipers and got a good look at his high-tech scope. They meant business! The Chinese also confiscated all photographic and communication devices during this period. In addition to the logistics and difficulty of the climb itself, this whole mess with the Chinese added to our concern.

Each day at Base Camp, we trained in the skills of ice-climbing. One of the very proficient Sherpas served as our trainer. This training also helped us bond with our team members, half of which were Sherpas. After 11 days at Base Camp, our whole team checked out well with 85 to 95 percent oxygen saturation. This meant that our acclimatization process had worked and our bodies were getting the oxygen we needed in spite of the altitude.

The south climbing route we were committed to crosses the infamous Khumbu Icefall. This treacherous glacial field of ever moving, toppling ice seracs (enormous columns of unstable ice) and countless crevasses is one of the most dangerous stretches of the climb. The Icefall has claimed many lives. Climbing Sherpas, fondly dubbed "Icefall

> *Christ has really changed George. He had a lot of faith through cancer. It was pretty nasty cancer and he made it through and recovered quickly and then climbed Everest! Without his faith and trust he may not have made it through those surgeries.*
>
> *He's been through a lot. His brother Bart died and then his sister Mia, and his mom. He's seen a lot of family members die. These are difficult times. Yet he is still standing strong in Christ.*
>
> *– Tim Whitworth, VP Investments, UBS Financial Services Inc.*

doctors," had gone ahead of us to set ladders and ropes in what they deemed the safest route across the Icefall.

We used the ladders as much as "bridges" over crevasses as we did to climb up and over sheer walls of ice. In spite of the ladders and ropes laid out by the Sherpas, the Icefall was still extremely dangerous. This is due to the fact that it moves as much as four to five feet per day. Experiencing the Khumbu Icefall as we did, we were duly impressed with the fact that Hillary and Tenzing crossed this Icefall under far more primitive conditions in 1953.

While sleeping at Base Camp, we could hear the Icefall just above us grinding and heaving all night. After May 31, the Icefall becomes too dangerous to negotiate because of the warmer weather and severe changes to its structure that the warm air brings. This is a major reason why the climbing season is so short on Everest.

At Base Camp we met a team of scientists from the National Aeronautics and Space Administration (NASA). They were studying the effects of low oxygen on the brain. One of the scientists, Jim Carter, requested that our team help them with their tests and research. At Base Camp, they submitted us to a series of aptitude and motor-skill tests. Throughout the climb, we retook these tests and performed various prescribed tasks numerous times at different altitudes to determine the effects of low oxygen on the brain. We reported and took these tests via special radios with which the NASA team had equipped us.

Mt. Everest, South Col Route

Base Camp – 17,500 feet (5350 m)
Camp 1 – 19,500 feet (5900 m)
Camp 2 – 21,500 feet (6500 m)
Camp 3 – 23,700 feet (7200 m)
Camp 4 – 26,300 feet (8000 m)
Summit – 29,035 feet (8850 m)

Even though all this testing and measuring was a substantial undertaking once we were on the mountain, we enjoyed being a part of their study. They also took blood samples and recorded other vital signs throughout the climb. NASA registered each of us on the Peak Freaks 2008 team as NASA Participants and gave us official NASA patches.

On April 15, the team prepared to climb to Camp 1 on the following day for the first time. The plan would be to sleep there the next night and then return to Base Camp. This is a 2,000 foot elevation gain and, of course, it would take us through the Khumbu Icefall. But high winds prevented our ascent to Camp 1 until April 21.

That evening, the team feasted on curry chick peas, lentil soup, and hot tea while the wind outside the tent blasted us. Amazingly, the tent held together!

On April 19, a Sherpa climber from another expedition collapsed in the Khumbu Icefall. This Sherpa had previously summited Everest twice, but the mountain is no respecter of persons. The Sherpa was evacuated by helicopter to Kathmandu.

The same day, we all witnessed an avalanche of monumental proportions as a large slab of glacier calved off the Col. This incident reminded us once again just how small and insignificant we were on this colossal mountain! Fortunately, Base Camp remained safe from this avalanche.

Strong winds are normal for Everest in the spring, as warm and cold air mingle. The warmer air will eventually usher in the monsoons, which will appear in form of heavy snow and poor visibility on the mountain.

On April 26, the weather service predicted winds of up to 120 km per hour on the summit. In order to complete a successful climb on Everest, everything is a balancing act: altitude acclimatization; food and water consumption; health; the elements; and in 2008 the Chinese!

Our plan was to continue our ascent to Camp 2 around April 22 or 23. Tim had shown himself to be an able guide and leader. He was full of energy, enthusiasm and knew his craft well. Yet he carried all this with humility— a rare, but comforting commodity in this environment.

We continued on up to Camp 2 on April 27 and spent a few nights up there before returning to Camp 1 and then back to Base Camp. As we were permitted, we climbed up and down between Base Camp, Camp 1 and Camp 2 continually.

Then, the Chinese insisted that the mountain be evacuated while they made their bid to the summit, so by April 30, our team was enjoying the richer air of Dingboche. There, we were able to communicate with our families and take a break from the mountain.

Meanwhile, we continued to wait on the Chinese. On April 29, Tim wrote:

> *Because no one is allowed into Tibet to give a report we have no confirmation on what the Chinese are actually doing. Rumors are that the torch is at Everest base camp on the Tibetan side of Everest and because they have asked that the Nepal side of Everest be closed May 1 to May 3 and the weather forecast is showing a couple days of calm, it is thought that they are planning on going to the summit during this time.*

It is odd that such a large international event would be kept so secret.[70]

As our Expedition Leader, Tim took the brunt of the bureaucratic assault on our team. Out of frustration, Tim wrote on May 4th:

George arrived a few days later than the rest of us. As soon as he arrived we were bombarded with statistics and facts regarding his chosen charity "America's missing children." Before too long, we were also able to recite the figures from memory due to the numerous occasions that we were reminded by George. I only mention this as it is indicative of George's energy and passion to promote and do everything he can do assist others, and in this case, a very worthy cause. In the short time I have known George I have never heard him say a bad word about anyone and he was possibly the most courteous and polite climber in Base Camp who would always go out of his way to help anyone. He has a heart of gold yet it is his bladder that is

Where are the Chinese? No one seems to know and there is much confusion at base camp in Nepal. One day you can make a sat phone call, the next you can't but someone else can. Yes, you can send out messages, no wait a minute—now you can't.

Organization of the rules at Base Camp seem to be playing out the same way. Everyone off the mountain—wait...no...okay maybe you can take some rope to Camp 3. Everyone not on the permit must leave at once...wait...well, okay you can stay.

What a season on Everest this has been. It has been hard to get organized when everything keeps changing. No matter, so far everyone is making the most of it and doing well and spirits are still

---

70 http://peakfreaks.com/everestnews2008.htm

Finally, we received word on May 8 that the Chinese had summited and were descending the mountain with the Olympic torch. With that news, military personnel left Base Camp as well, thanking us for our hospitality and cooperation. This meant that we could now make plans to proceed to Camps 3 and 4 and then make our bid for the summit!

That night it was as if the mountain groaned in travail. Tim wrote:

> Last night was a very noisy night. The mountains were alive train-wrecking all around us. Big chunks were slamming down off Mt. Pumori and the Icefall was crashing and groaning. The days are getting warmer which will start to deteriorate the structure of the Icefall. Have no fear though because the Icefall doctors have built the route away from any of the dangerous areas and will be maintaining the ladders throughout the day.[72]

high, all of which is most important at the end of the day.[71]

more impressive as we were all amazed at George's incredible ability to fill a 4 liter bottle with urine each night!! (The pee bottle phenomenon is also the subject of much discussion at base camp). Finally, I feel that it is my duty to inform his family that George is now addicted to the series "24" with Kiefer Sutherland. Every night, when George rose from the dinner table and placed himself in front of the DVD I knew that it was time to set up for the evening's episode. George's family would be well advised to buy a few of the box sets of "24" to assist George on his return to civilization.

– Nabil Lodey

On our treks through the Khumbu Icefall, whenever we came to a horizontal metal ladder spanning a deep crevasse, we had to cross it like a bridge. Because we wore metal crampons

---

71 http://peakfreaks.com/everestnews2008.htm
72 http://peakfreaks.com/everestnews2008.htm.

strapped to the soles of our boots, we were stepping metal on metal. This often proved very slippery and unstable.

On one descent climb through the Khumbu Icefall I fell into a crevasse. My foot became lodged in a small crack and my body twisted around until I was hanging upside down with all my weight on my ankle! The fall severely twisted my ankle and I heard and felt a snap, crack and pop. I was sure I had broken it. Years before this I had broken my ankle in a skiing accident and had two pins put in, so I knew what a break felt like. With my Sherpa's help, I managed to contort my body and free my ankle. Hobbling back to Base Camp, I thought for sure I had broken my ankle and this was the end of the climb for me.

That night my ankle swelled up to the size of a grapefruit. I stayed up all night putting hot and cold packs on it and praying. I prayed, "God, I know You brought me this far and if I'm not to summit Everest then so be it. But I would sure like to finish this climb and I know You can heal me if You choose to do so."

As I prayed and cared for my ankle all night, I meditated on the Scripture in Matthew 9:6-7: "Then Jesus turned to the paralyzed man and said, 'Stand up, pick up your mat, and go home!' And the man jumped up and went home!" I knew that Jesus could heal me like He had healed that paralyzed man, but was it His sovereign will to do so? I wouldn't know until morning when I could have my ankle examined by a doctor.

Chapter Twenty-Seven
— THE SUMMIT! —

The true summit is getting back to base.
Anonymous

THE NEXT MORNING I got up and "Thank God!" the swelling in my ankle was gone! I had already resigned myself to the fact that I probably wouldn't be able to finish the climb, but God performed a miracle. I put my boots on (even the fact that I could get my boot on was incredible) and went over to the medical tent to have them check out my ankle. I didn't want to put my team or me in jeopardy if I was still injured. The medical team examined my ankle and couldn't find anything wrong with it. They pronounced me fit for the climb. I walked back to my tent thanking Jesus for healing me!

By mid-May, we had been on Everest for nearly 50 days. Tim was constantly checking in with each of us to monitor how we were doing. On May 14, he wrote the following journal entry:

At this stage of the climb the climbers are starting to show the effects of living at altitude. They lose incredible amounts of weight at altitude. It has been seven weeks now and a week or more to go before the summit push. They have by now lost all fat reserves and even continue to do so while resting. In this harsh environment the body will start to consume muscle when there is no fat left.

It is always such a balancing act when climbing Everest. Acclimatization with health, time needed to recover, trips up and down to avoid AMS, hauling loads, sitting out weather. The down time with the Chinese invasion didn't help matters adding another consideration in the equation. Thankfully the weather isn't hammering them too bad this year.[73]

At high altitude, we were burning an estimated 6,000 calories per day. The problem with this is that the low oxygen had the effect of reducing our appetites and ability to digest the food we needed. In order to compensate for this, we had to be very disciplined about what we ate, how much and when. Tim and his team had this well-thought out and well-planned. I remember one delicious supper in which we enjoyed fresh organic roasted chicken, potatoes, greens and banana pie.

Again, my motto on Everest was to "drink like a fish and eat like a pig." This is the recommendation I give to anyone who is considering a high-altitude climb, in addition to making their reservations with Jesus Christ.

During the days of May 11 through 17, we climbed from Camp 1 to Camp 2, slept there two nights; then we proceeded to Camp 3, where we slept one night. The following day we retreated back down to Camp 2 to rest.

On May 17, Tim announced the exciting news that tomorrow we would begin our push to the summit! We were all very excited at the announcement! We slept that night at

---

73 http://peakfreaks.com/everestnews2008.htm.

Camp 2 and then climbed to Camp 3 on the 18th. Unfortunately, Nabil had caught some kind of bug and we had to leave him at Base Camp. He would not be summiting with us. We slept that night and the following night at Camp 3 with a higher altitude climb on the 19th to continue the acclimatization process. By the end of the day, we were all feeling great.

Then, on May 20 we made the seven-hour climb to Camp 4. Camp 4 put us in the *death zone*. The death zone is any altitude above 26,000 feet (7880 m). At this altitude and above most humans lose all ability to acclimate. As a result, the body begins to slowly deteriorate and die. Cuts and bruises do not heal; a cold or other illness won't go away.

Reaching Camp 4, the team rested for another six hours. But I came in late because I was urging Sultan on. He was a solid climber and my friend, but he was very slow. This meant that he and I only got two hours of rest before we made our summit bid. It seemed as though I had just laid my head down to sleep when I heard Tim waking us up, "We gotta go! We gotta go!" Due to numerous factors, not the least of which was our safety, we were on a strict schedule to summit and return to Camp 4 the same day.

With no more than two hours of sleep, I was taxed out! We had stripped everything out of our packs that we didn't absolutely need for the summit climb. In our rush to leave and in

> *George is the consummate entrepreneur! There are no closed doors in his life. Every door is unlocked. There's no mountain he won't climb.*
>
> *George called me from the summit of Everest. Here's a guy dying from cancer and he climbs Everest and helps save a guy's life! George wasn't even a mountain climber!*
>
> *– Patrick McCourt, Entrepreneur and friend*

my muddled state of mind, I inadvertently left my headlamp in my tent, an oversight I would later regret.

We started for the summit that evening at 10 pm. As we left Camp 4, the night was clear with a full moon, so it was easy to see our way. The view before us was incredibly beautiful—mystically so. We joined an already long line of other climbers, all pushing for the summit. We were blessed to have good weather.

Tim had ensured our success to the extent humanly possible with his rigorous schedule for acclimatization, planning our ascent from Camp 4 instead of from Base Camp, having 85 bottles of oxygen in place, and recruiting a total of 11 climbing Sherpas to team up with us.

In the early hours of May 21, we made the South Summit. This was a major goal for us and the turn-around point if the weather had deteriorated. The South Summit is a dome of snow and ice at 28,700 feet about the size of a ping pong table. From the South Summit we could now see the remaining landmarks between us and the summit: the Cornice Ridge, Hillary Step and the final grade to the summit. Peak Freaks offer a more detailed description of these features:

> *Who has cancer and then climbs Mt. Everest three months later?*
>
> *I was in Miami, Florida having a martini at a restaurant with a friend and got a phone call. It was George calling me from the summit of Mt. Everest on a sat-phone. That was so cool!*
>
> *Consider what George has done raising funds for ACIM without any training or background. Who does stuff like that? He'll try anything. If he wakes up some morning and decides to be president, don't be surprised if he makes it!*
>
> *– Joe Esquivel, CFO and friend*

> *The Cornice Ridge is a 400-foot long horizontal section of rock and wind-carved snow, this is easily the*

most intimidating section of the climb. Climbers must carefully traverse a knife-edge ridge of snow plastered to intermittent rocks. This is the most exposed section of the entire climb, and a misstep to the right would send climber tumbling down the 10,000-foot Kangsung Face. A misstep to the left would send one careening 8,000 feet down the Southwest Face, were it not for the fixed ropes.

At 28,750 feet, the Hillary Step is the most famous physical feature on Everest. The Hillary Step is a 40-foot spur of snow and ice. First climbed in 1953 by Edmund Hillary and Tenzing Norgay, the Hillary Step is the last obstacle barring access to the gently angled summit slope. Modern-day climbers use a fixed rope up here to ascend the Hillary Step. We marvel at Hillary and Tenzing's achievement in climbing this impressive mountaineering obstacle without fixed ropes and using what is now considered primitive ice climbing equipment.[74]

As our team pushed toward the summit, we were quite strung out based on individual speed and other factors. The crowd of climbers was thick on the mountain and slowed everything down, causing us to use up more oxygen than we had hoped.

At one point, it may have been at the Cornice Ridge, the group of climbers in front of us had come to a complete standstill. We tried to determine the holdup, but it appeared that they were simply frozen with fear due to the steepness of this section

---

74 http://peakfreaks.com/everestnews2008.htm.

and its extreme exposure. We needed to keep going, so I simply called out, "Coming through!" and we went around them. We didn't want to die because of someone else's fear or stupidity.

Dom was the first of our team to summit at 7am, followed by Larry at 11:30am, Scott at 12:30pm, and Tim and I shortly after 1pm. Farouq was still ascending as we made our descent.

So it was that I summited Mt. Everest on May 21, 2008, at 1:08 pm Nepal time, standing at 29,035 feet.[75] Six of our Peak Freaks team: Dom, Larry, Scott, Farouq, Tim and I; and seven of our Sherpas: Mingma Sherpa, Lhakpa Bhote, Gelgan Sherpa, Karma Sherpa, Dendi Sherpa, Kajee Sherpa, and Ang Pasang Sherpa made the summit. On the summit we shot some photos, we recorded a short message for ACIM, and I made several quick satellite phone calls to my family, friends and sponsors.

Unfortunately, three members of our Peak Freaks team were unable to summit: Nabil we had left in Base Camp due to illness. Saad was forced to abandon his attempt at a speed ascent when he slipped on a ladder in the Icefall early in his climb and injured his leg. Finally, Sultan ran into trouble on the ascent and couldn't summit. We had really become close as a team and I wanted to see all these guys make it, but it was not to be.

Reaching the summit is so exhausting and the climb back down so perilous that we didn't take much time to enjoy the view. Getting back to Camp 4 became our number one goal. Often the trip down is more dangerous than the climb up. As

---

75 The exact height of Mt. Everest has long been disputed. According to the US National Geographic Society, its height is 29,035 feet determined by GPS technology. http://news.discovery.com/adventure/activities/everest-official-height-120301.htm. See also: http://www.britannica.com/EBchecked/topic/1673089/Height-of-Mount-Everest.

veteran climbers say, "The true summit is getting back to base." Put another way, Ed Viesturs, the world-class American mountaineer warns: "Reaching the summit is optional. Getting down is mandatory."[76]

In fact, most deaths on Everest occur during the descent on the upper slopes in the "death zone." "The high elevation and corresponding lack of oxygen coupled with extreme temperatures and weather conspire to create a greater risk of death than on the mountain's lower slopes."[77] Not to mention the fact that you're operating on very little sleep and have spent your energy reaching the summit.

Scott Mortenson wrote, "For me, summiting Mt. Everest was akin to running up to the penthouse of a burning building, grabbing your precious photos and then trying to make it out alive. Yet, everyone was moving so slow! An uneasy feeling was growing in my stomach...."[78]

The *death zone* was like an invisible specter looming over us waiting to pounce and claim its prey. And several of us nearly succumbed to this specter's chilling devices.

---

[76] Ed Viesturs, *The Mountain: My Time on Everest*, (New York: Simon & Schuster, 2013), p. 58.
[77] Stewart Green, "Death on Mount Everest," www.Climbing.About.com., 2012.
[78] http://peakfreaks.com/everestnews2008.htm.

Chapter Twenty-Eight
— DISTRESS CALL —

It was titillating to brush up against the enigma of mortality, to steal a glimpse across its forbidden frontier. Climbing was a magnificent activity, I firmly believed, not in spite of the inherent perils, but precisely because of them.
Jon Krakauer, author of *Into Thin Air*

ONLY A SHORT while after we had summited, Tim received a distress call from Sultan who was still below us saying that he was dehydrated. Tim told him to ask another climber for water and told him we'd bring some to him as quickly as we could get there. We had taken less than half an hour on the summit. We nearly flew down the mountain from the summit to come to Sultan's aid. Tim, Scott, Larry and I reached Sultan at the South Summit at 2 pm. He was sitting down and refusing to move. He was delirious, uncooperative and began losing consciousness.

At one point, Sultan lost consciousness and stopped breathing. We thought he had died. Scott wrote:

> *I was about to deliver two rescue breaths and begin CPR when Tim tried the old school precordial thump—a hard fist to the chest which is a desperate attempt to provide enough stimulus to get the heart beating again. Whatever the physiological reaction was, Sultan inhaled again and though he denies ever*

losing consciousness it was clear to me that we were still in serious trouble. Successfully rescuing a patient from this high in the Death Zone was a rarity in the annals of mountaineering—especially, when the would-be rescuers were exhausted from a grueling two days of climbing.[79]

Tim, Scott, Larry and I immediately went into action, cutting and tying ropes to create a rope rescue system with which we could lower Sultan down the mountain. Tim knew we were going to need help to get Sultan down alive. We had all been climbing at high altitude all night and day and were physically spent. Scott was just about out of oxygen, so Tim sent him to Camp 4 to get help and emergency supplies. Tim asked Larry to give Sultan his oxygen bottle and to descend to Camp 4 as well.

Soon thereafter, Tim turned to me and said, "You need to get going, George." I had paid for my own personal Sherpa (Dendi Sherpa) and extra oxygen and so I asked Dendi to stay with Tim and help save Sultan's life. This was without question a life-saving decision, though also somewhat cavalier because I thought I could descend by myself. Coming down is the most dangerous part of the climb and I was physically exhausted, low on oxygen, without a headlamp and now without my Sherpa team mate.

A solo descent on Everest is not wise, but it was the choice I made to help save Sultan's life. Without my Sherpa to assist him, there was no way Tim could've gotten Sultan down as far as he did. Meanwhile, Tim was counting on fresh Sherpas coming up from Camp 4 to help with Sultan's rescue.

---

79   http://peakfreaks.com/everestnews2008.htm.

As if Tim didn't already have enough on his hands, he and those with him came upon another climber in distress. Tim writes,

> As we were working our way down, we came upon a climber in distress - an older Korean fellow sitting on the route, stuck in old fixed rope, with the new fixed rope pinching his shoulder in a position such that he couldn't move. I managed to cut the old rope under him, move the rope from his shoulder that had trapped him, and slid him down and to one side. He was obviously terrified. When I moved him, I discovered he had been sitting a puddle of his own urine. Poor fellow, I couldn't help him much more than that as my hands were full with Sultan. I did however have him and his pack now properly hooked up to the new fixed line and he managed to scoot down on his butt, and eventually showed up at C4 (Camp 4).[80]

Scott sheds some light on how precarious our situation had become,

> "Tim, I want to get the hell off this mountain." I said after another lengthy delay waiting for a man ahead of us to take ten minutes to swing his leg over a rock. My sense of unease was now an absolute feeling of impending doom. "Me too." Tim said. "Me too."

As I continued my descent, I was running very low on oxygen. Oxygen deprivation makes you feel delirious, perhaps a bit like being drunk or not in full control of your faculties. You try to focus and pierce through that. You have to consciously overcompensate for the feeling of confusion.

---

80   http://peakfreaks.com/everestnews2008.htm.

By nightfall I came across Larry lying face down in the snow. The wind had picked up and it was snowing heavily. The temperature had dropped to below zero. The snow was covering him and he appeared to be in bad shape. If he didn't get up immediately we would lose him. I yelled at him trying to make myself heard over the blast of the wind and snow, "Larry, what are you you doing?" He mumbled, "George, get help." "Larry, by the time I get back with help, you'll be dead! You need to get up right now and share my oxygen with me."

With that I jerked Larry up off the ground. The fact that Larry's mouth was covered with white foam told me he was in the throes of death. I pulled him up, wiped off his face and we shared my oxygen—what little was left. (We were all running low on oxygen because of our extended stay on the mountain trying to help Sultan.) Had I not come across Larry when I did, he no doubt would have died.

As we continued our descent, we searched for a reserve oxygen tank for Larry, but couldn't find one that was compatible with his unit. There were so many different nationalities of climbers on the mountain and no standardized fittings. Finally, we found an oxygen bottle. But as soon as Larry had fresh oxygen, he bolted from me leaving me alone on the mountain! Between the two of us, he had the only head lamp. In our hurry to leave Camp 4, running on little sleep, I had forgotten my headlamp. Cognitive faculties are not what they should be when deprived of oxygen. Add to that our extreme exhaustion and this explains our befuddled behavior on the mountain.

Later that night I ran out of oxygen and at one point slipped in the dark. I fell and was hanging upside down on the mountain by my mask around my neck. My mask was caught on something and I was hanging from my mask like a hangman's noose nearly choking, but finally was able to free myself. Even though climbers from other teams passed close by me in this condition, no one stopped to help. There's no hand-holding on Everest. The attitude of many is, "If I help you, it may cost me my life."

I had released myself from the tangle and continued my descent. At this hour of the night, the south side of the mountain was shielded from moonlight, so it was pitch dark. Unable to see my way, I fell several times tearing open my down suit on the sharp rocks.

This late in May, we were nearing the end of the climbing season and there were many climbing ropes set up on Everest. We would clip into these climbing ropes either with a Jumar ascender for the ascent, or a carabineer on the descent to prevent a fatal fall. (In the 2015 movie, *Everest*, Doug Hansen failed to clip into one of these ropes properly and he simply stepped off the mountain to his death.) One's befuddled state of mind and the extreme cold made it difficult to clip into these ropes.

Added to that, these ropes had been set and reset numerous times, and some of them were no longer anchored to anything. In total darkness, I had to make sure that each rope I clipped into was actually secured. So each time I found a rope, I grabbed hold of it and yanked on it vigorously to ensure it was secure before clipping into it. Toward the end of my descent to Camp 4, a climber from another team caught up with me and shined his headlamp ahead of me so I could see.

In the 2015 movie, *Everest,* Rob Hall's team stayed in constant communication with their Base Camp manager. In our case, Tim Ripple was our mobile Base Camp manager and expedition guide. We had no one at Base Camp with whom we stayed in contact. As he could, Tim kept his wife Becky (who was back in Canada) appraised of our situation. While I was still making my descent, Scott had made good time getting down to Camp 4. He called out among the tents seeking help for Sultan, but got no volunteers. Several times he tried to offer money to anyone who would assist in Sultan's rescue, twice increasing the amount, but no one would come forward. Finally Scott found a Sherpa from another team who was going up early the next morning and asked if he would also take a sleeping bag and stove for Sultan. We needed to prepare Sultan for another night out on the mountain.

Tim fills in what was going on with him as he continued efforts to rescue Sultan higher up on the mountain, still in the death zone:

All Sultan wanted to do is sleep. We were now prepared to let him do exactly that. There was no way we could physically carry him. At this point, Sultan had become violent and a danger to his rescue party. At one point he heaved a rock at me, hitting me square in the forehead, knocking me off balance, and almost sending me down the South Face. Luckily I was stopped by a Sherpa on his way up.

The Sherpas were no longer interested in the events that were taking place and I respected that. We had been working extremely hard to lower and drag Sultan; it was now over 32 hours we'd been up there, most of it without oxygen. We were dragging Sultan and would pendulum him through the steep pitches. The rocky flat sections were very difficult having to semi-carry him and drag him, trying not to rip open his down suit.[81]

By 3 am on May 22, the Sherpa that Scott had recruited delivered the sleeping bag to Tim and Sultan. Sultan was still above Camp 4. While all this was going on, we learned that Farouq was suffering from snow blindness! He had been "showboating" on top of Everest, posing for many pictures without his goggles on. The result was that reflective sunlight off the snow had blinded him. A Sherpa was assisting him in his descent while we continued to rescue Sultan.

Tim writes:

We put Sultan in the sleeping bag and tied him off on a ledge so he wouldn't roll off, straight across from Scott Fisher's dead body, hoping Fisher might tell him to get out of there, that was his place, go home! From what happened next, maybe he did?

I managed to get some water and food in Sultan. I tried to get him to take the oxygen but he kept throwing it off. I covered his face with his Oman flag, tucking it in around his hood to protect it from the elements. I told him I would be back in a couple hours - we were much closer to Camp 4 now so I figured I could quite easily make it back up in good time to take over where we left off, after we rested. We decided to go at once and get some rest for all of us, warm our feet, and get more provisions.

---

81 http://peakfreaks.com/everestnews2008.htm.

It was around 4:30 am I retreated to C4, with the Sherpas, and we crawled into our tents. It was just a short time later a Sherpa came down without actually looking at Sultan reporting that Sultan was dead. Willie passed this news on to me. Needless to say, I was devastated. I had just left him, how could it be? I was in shock. The feeling inside me was something I had never felt before. 27 years guiding and I have never lost a client, it didn't seem real.

Around 7:00 am I heard someone yelling, "Tim, Tim, Sultan is alive!" I looked out of my tent to see two Sherpas from the Indian Army escorting him down. I couldn't see him at first because I was looking for someone carrying him down but then my eyes focused in on his familiar boots, he was walking! Apparently the Indian army Sherpas passed by him and saw him rustling in his sleeping bag. He was wide-awake now and somewhat refreshed, got up and walked down to the South Col. This was music to my ears, not only did he come back from the dead but he walked down too and I didn't have to go back up and get him.[82]

On the following day, May 23, my Sherpa Dendi and I escorted Sultan to Camp 3 and we stayed with him overnight to ensure he survived. Tim had been able to assist Farouq down to Camp 2 and was waiting for his snow blindness to heal before they would attempt to negotiate the Khumbu Icefall. The morning of May 24, the three of us continued our descent to Base Camp. I was the last person from our team on the mountain. I wanted to make sure that no one was left behind.

---

82   http://peakfreaks.com/everestnews2008.htm.

In Base Camp, I gave most of my gear and any leftover money to our Sherpas. I gave my thousand-dollar climbing boots to Dendi Sherpa. There was no way they could afford the kind of gear we had and it seemed only right to pass it along to them.

In view of all that had transpired up in the death zone, what a relief that by May 24 the whole Peak Freaks team had made it back to Base Camp! It took us a few days to trek out to Lukla and catch a plane back to Kathmandu. I attribute the key to my success on Everest to God as I prayed every step up and down the mountain.

We admitted Sultan into the hospital in Kathmandu and I stayed with him while they checked him over. He appeared to be healthy except for two or three finger tips he may have lost due to frostbite. While at the hospital, I took the opportunity to weigh myself and discovered that I had lost 37 pounds on Everest!

After seeing to Sultan's welfare, my stomach gnawed with a voracious hunger! As crazy as it sounds, what I yearned for more than anything else was hot apple pie ala mode. I trekked all over Katmandu until I finally found a restaurant that served this all-American dessert. I savored every bite!

The fact that Dad climbed McKinnley and Everest is understated. Those feats are absolutely spectacular! He doesn't realize how amazing these accomplishments were.

I'm impressed by his ability to go after any project no matter how big. He's never scared of or intimidated by anything! He reaches for the stars. My dad always assures me, "You can do anything you want!"

I've started singing and have produced my first album. My dad is my biggest supporter behind my singing. He always believes in me. He's my biggest fan.

– Ashley LaMoureaux, daughter

On the long trip home, I had time to reflect on the past three months. I felt relieved to have been able to return safely from the climb without losing anyone on our team. I also recognized how fortunate I was to have been able to summit Everest. This had been Tim Ripple's seventh attempt, but only his first summit. I was truly blessed to have summited on my first attempt.

I also recognized that without the motivational help of Dr. Kenneth Friendly's message *Never Quit* and without the nutrient-rich food prepared for me by Steve Planté and his Organic Oasis, I could never have accomplished the climb.

God was truly good to me allowing me to climb the tallest mountain on earth just two-and-a-half months after five, back-to-back cancer surgeries. I really wanted my story to inspire others to not give up when they lose their health, family, or fortune.

I was also excited about continuing to promote the work of ACIM and how this climb might move that important work forward.

Black Belt newspaper article: "Anchorage Daily News announcement of George receiving his 1st Degree Black Belt in Karate"

George peeking out of tent: "George on Mt. Everest preparing for the ultimate climb!"

George W. Bush,
President of the United States of America

Mr. LaMoureaux,

The President has asked me to thank you for thinking of him and to convey his best wishes for a successful event. [Climb for America's Children]

Bradley A. Blakeman
Deputy Assistant to the President

Governor William J. Sheffield
3125 Susitna View Court
Anchorage, Alaska 99517

December 18, 2001

Climb for America's Children
Attn: George LaMoureaux, Founder
3705 Arctic Blvd., #1124
Anchorage, Alaska 99503

Dear George,

I appreciate you briefing me on the Climb for America's Children event and A Child Is Missing programs which are so very important to this country. As you explained, the program was founded by Sherry Friedlander and is expanding nationwide, Alaska being one of the states that will benefit from its service.

You will make Alaska proud that you have come up with this interesting way to assist in raising funds to not only bring the program to Alaska but to help other states and the children in those states. The awareness to the public that your climb will generate will alert the families in their time of need that there is immediate help for them to find their child that is missing in the first hours of their disappearance.

There are over 3,000 missing children reported daily in our country; this is incredible. Speeding up the search process through A Child Is Missing rapid response system to the area they were last seen and alerting the people in the area of the child's disappearance will assist law enforcement in the recovery of these missing children.

This is a tremendous undertaking and I congratulate you and your team that are willing to use their experience in helping in this important program.

George, your experience of successfully building companies, putting things together, working with a vision and knowing lots of people in business and the media makes a big difference in an undertaking such as the Climb for America's Children. You have my support.

Sincerely,

Bill Sheffield

William J Sheffield
Former Governor of Alaska

December 30, 2001

Mr. George LaMoureaux, Founder
And National Advisory Board
Member of a Child Is Missing
3705 Arctic Blvd. # 1124

Dear George,

Programs, which benefit children in need and reunite them with their families, deserve the support of all. The Mountaineering Club of Alaska, with its special purpose of promoting the enjoyment of hiking, climbing and exploration of the mountains in a safe and environmentally friendly manner, gives its support to the "Climb for America's Children" so that children can be reunited with their families… and they too can enjoy the "freedom of the hills."

                                        Sincerely,

                                       Richard Baranow
                                       Mountaineering Club of Alaska

TONY KNOWLES
Governor
governor@gov.state.ak.us

STATE OF ALASKA
OFFICE OF THE GOVENOR
JUNEAU

P.O. Box 110001
Juneau, Alaska 99511-0001
(907) 465-3500
FAX (907) 465-3532
www.gov.state.ak.us

January 23, 2002

Mr. George LaMoureaux, Founder
Climb for America's Children
3705 Arctic Boulevard, No. 1124
Anchorage, Alaska 99503

Dear Mr. LaMoureaux:

Please accept my warmest welcome to the Climb for America's Children Denali 2001-2002 Expedition, a benefit for the nonprofit organization, "A Child Is Missing." You could not have chosen a more spectacular state for "The NeXt Generation Telethon" and sporting event.

I applaud your efforts in organizing this extreme sport charitable event to raise funds for the National Command Center. Heightened awareness of the "National Rapid Response System" will enable law enforcement officials to find missing children across the country, a truly worthy cause. Your selfless commitment to the success and future development of "A Child Is Missing" is extremely impressive.

On behalf of all Alaskans, thank you for your dedication to serving our country and communities in the past, present, and future. I wish you the best of luck on a successful telethon and sport event!

Sincerely,

Tony Knowles,
Governor

MARK FOLEY
16TH DISTRICT, FLORIDA
DEPUTY MAJORITY LEADER
WAYS AND MEANS
COMMITTEE
SUBCOMMITTEE ON OVERSIGHT
SUBCOMMITTEE ON SELECT AGENDA
MEASURES

Congress of the United States
House of Representatives
Washington, DC 20515

REPLY TO
104 CANNON BUILDING
WASHINGTON DC 20515-0916
(202) 225-5792
FAX: (202) 225-3132

E-MAIL: markfoley@mail.house.gov
WEBSITE: http//www.house.govfoley

February 4, 2002

Mr. George LaMoureaux, Founder
Climb for America's Children
3705 Arctic Boulevard, No. 1124
Anchorage, Alaska 99503

Dear Mr. LaMoureaux:

I would like to applaud you for founding The Climb for America's Children event. You and your program have taken on the daunting task of finding missing children, which is one of the toughest and most challenging goals imaginable.

Every day that passes, more than 3,000 children are reported missing, and most Americans have no idea this is even occurring. It is always an emotional and physical drain that takes an enormous amount of commitment, dedication and faith to keep going for everyone involved.

As Co-Founder of the Congressional Missing and Exploited Children's Caucus, and Chairman of the Entertainment Task Force, I want you to know you'll have my full support in meeting the challenges you face. I can't think of a more worthy cause, and I know it will be a huge success.

Sincerely

Mark A. Foley
Member of Congress

PALM BEACH GARDENS
4480 PGA BLVD, SUITE 406
PALM BEACH GARDENS, FL 33410
(561) 627-6192
FAX (561) 626-4749

PORT ST. LUCE
COUNTY ANNEX BUILDING
290 NW COUNTRY CLUB DRIVE
PORT ST. LUCE, FL 34306

HIGHLANDS COUNTY
SEBRING CITY HALL
(By Appointment Only)
(563) 475-1813

Walter J. Hickel

P.O. Box 101700
Anchorage, Alaska 99510-1700
Tel 907-343-2400
Fax 907-343-2244
E-mail: wjhickel@gci.net

February 11, 2002

Mr. George LaMoureaux, Founder
Climb for America's Children, and
National Advisory Board Member of "A Child Is Missing"
3705 Arctic Boulevard #1124
Anchorage, AK 99503

Dear Mr. LaMoureaux,

Congratulations on being named to the National Advisory Board of A Child Is Missing.

I was astounded to learn that over 3,000 children a day are reported missing in America, according to FBI figures, and last year 4,241 missing children cases were in Alaska alone. Even when most children are found, it is a traumatic experience. Something this horrifying needs to be dealt with in an immediate and strategic manner.

Your support for the National Command Center and the National Rapid Response System (now nationwide because of your fundraising campaign) has enabled and will continue to help law enforcement officials to find missing children here in Alaska and across the nation. This is a noble cause that needs broad support.

Good Luck with your "Climb for America's Children" charitable event here and the "A Child Is Missing" program.

Sincerely,

Walter J Hickel

Walter J Hickel
Former Governor of Alaska
Former Secretary of the Interior

February 13, 2002

Mr. George LaMoureaux, Founder
Climb for America's Children
3705 Arctic Boulevard, No. 1124
Anchorage, AK 99503

Dear Mr. LaMoureaux,

On behalf of the American Mountain Guides Association, I would like to express our support for "Climb for America's Children."

The American Mountain Guides Association stands behind your endeavors to raise awareness about missing children in our country. We believe you have come up with a creative way to bring this problem to the attention of the American people through your "NeXt Generation Telethon and Extreme Sport Charity Event."

We are excited to see your organization facilitating change through the sport of climbing. We applaud your efforts and wish you the best of luck for a successful Denali Expedition and charity event.

Sincerely,

Simon Fryer,
Program Manager
American Mountain Guides Association

JEB BUSH
GOVERNOR OF THE STATE OF FLORIDA

February 15, 2002

Dear Mr. LaMoureaux:

It is with great pleasure that I recognize you for the found of the Climb for America's Children event.

More than 3,000 children are reported missing every day. We need to increase public understanding and awareness of this problem to prevent this alarming rate from continuing. I appreciate your commitment to your community and the contributions you have made to enhance both our way of life and our society as a whole.

Congratulations on all your accomplishments and my warmest greetings and best wishes on your continuing success.

Sincerely,

Jeb Bush

Chapter Twenty-Nine
— GERSON THERAPY —

Let food be thy medicine and medicine be thy food.
Hippocrates

AFTER RETURNING TO the States, I turned my attention to discover the cause of my sister Mia's death. My family and I were shattered by her death and the circumstances surrounding it. The newspaper ran three articles, the first following the day of her death:

> *Police are investigating as a homicide the mysterious death of a 48-year-old Anchorage woman. Mia Soltis, a dancer at the Crazy Horse, was found dead on the deck of a Jewel Lake home Sunday afternoon, one day after her purse had been found by a man walking his dog in the neighborhood and a few hours after her father had reported her missing. She was last seen early Saturday morning at work, police said.[83]*

A week and a half later, the paper printed an update:

> *Ten days after Mia Soltis' body was found on the deck of a Jewel Lake home, Anchorage police still don't know how she died.[84]*

---

83 Beth Bragg, "Dancer's Body Found at Jewel Lake Home," Anchorage Daily News, March 31, 2008, p. A3.

84 James Halpin, "Cause of Woman's Death Still Unknown," Anchorage Daily News, April 9, 2008, p. A3.

Mia was neither a heavy drinker nor a drug user. She was extremely athletic and fit. Only recently had she started dancing at the Crazy Horse at the urging of her husband. I discovered that a co-worker of Mia was murdered the same night Mia died just a few blocks away from where Mia's body was found. My belief is that Mia witnessed the murder and ran for her life. Neither her purse nor her shoes were with her body. March in Anchorage is still winter and no time to be without shoes.

Unable to solve the case, the police labeled it a "misadventure" and sealed her file for the next 50 years. I asked them, "What the hell is a *misadventure* and why was her file sealed for 50 years?" In my opinion, there were some shady things going on and someday I hope to find out what happened and bring to justice whoever is responsible for her death. The whole thing was very suspicious and mysterious. Mia was survived by her husband Patrick and her two girls fourteen and five years old.

Five months after my return from Everest, I went back to Virginia Mason Hospital in Seattle for a follow up visit. They injected me with radio-active isotopes and I lit up indicating the presence of cancer, but it was just a small spot. The cancer was in my neck near one of my glands. So they took me in for my sixth cancer surgery. This was just eight months following my fifth surgery and right after being on Everest for 77 days. I had really hoped I was through with cancer, so this came as quite a blow. But I know the Everest climb had taken a lot out of me. Also, had I known about the Gerson Therapy before my surgeries, I would not have endured any of those surgeries, but would have chosen Gerson Therapy instead.

After my surgery, I consulted with doctors at both Virginia Mason Hospital and the Swedish Cancer Institute in Seattle.

Doctors at both hospitals recommended radiation and chemo-therapy. As a part of my surgery, the radiation specialists had prepared a restraining device that was contoured to my body. This device would prevent me from moving during the radiation treatments. The doctors unanimously insisted I go through both radiation and chemo-therapy, otherwise they were sure the cancer would kill me. At the time, I just figured this is what I have to go through to get rid of the cancer.

However, the doctors explained later that the radiation therapy would be very destructive to my muscles, teeth and bones. They warned that the radiation would permanently damage my saliva glands and that the inside of my mouth and throat would feel always feel dry like a piece of over-cooked chicken. I'd have to carry a water bottle around with me for the rest of my life to constantly dampen my parched mouth and throat. In short, my quality of life would be gone.

I wasn't afraid of the treatment; I simply decided it wasn't for me. While in Seattle, people from the Cedar Park Assembly of God Church prayed over me several times, asking God to heal me.

In a way similar to that of launching a business in a field about which I knew little, I devoted myself to studying cancer, sifting through thousands of pages of research on cancer therapy. This was a major event and I was sweating it out, afraid that I wouldn't be around much longer to take care of my daughters Ashley and Tamra.

I began contacting numerous clinics for alternative treatments for cancer. My former wife, Carol, recommended the Gerson Institute (www.gerson.org) and the Gerson Therapy. Carol's dentist had experienced the Gerson Therapy and had been healed of his cancer. While the Gerson Clinic is in Mexico, I found the Gerson Institute in Los Angeles. I opted to go to Los Angeles instead of Mexico for my treatments. Of course, in the US all forms of cancer therapy other than surgery, radiation, and chemotherapy, or experimental pharmaceuticals are illegal.

Even so...

> *The Gerson Therapy is a natural treatment that activates the body's extraordinary ability to heal itself through an organic, vegetarian diet, raw juices, coffee enemas and natural supplements.*
>
> *With its whole-body approach to healing, the Gerson Therapy naturally reactivates your body's magnificent ability to heal itself – with no damaging side effects. This a powerful, natural treatment that boosts the body's own immune system to heal cancer, arthritis, heart disease, allergies, and many other degenerative diseases. Dr. Max Gerson developed the Gerson Therapy in the 1930s, initially as a treatment for his own debilitating migraines, and eventually as a treatment for degenerative diseases such as skin tuberculosis, diabetes and, most famously, cancer.*
>
> *The Gerson Therapy's all-encompassing nature sets it apart from most other treatment methods. The Gerson Therapy effectively treats a wide range of different ailments because it restores the body's incredible ability to heal itself. Rather than treating only the symptoms of a particular disease, the Gerson Therapy treats the*

causes of most degenerative diseases: toxicity and nutritional deficiency (emphasis mine).

An abundance of nutrients from copious amounts of fresh, organic juices are consumed every day, providing your body with a super-dose of enzymes, minerals and nutrients. These substances then break down diseased tissue in the body, while coffee enemas aid in eliminating toxins from the liver.

Throughout our lives our bodies are being filled with a variety of carcinogens and toxic pollutants. These toxins reach us through the air we breathe, the food we eat, the medicines we take and the water we drink. The Gerson Therapy's intensive detoxification regimen eliminates these toxins from the body, so that true healing can begin.[85]

I flew to Los Angeles with my mom. For about three weeks, I submitted myself to the regimen of the Gerson Therapy. This involved an organic juice diet, coffee enemas, and taking many different supplements. In addition to the above, they explained that they were alkalizing my body to suppress the growth of cancer. As Steve Planté had shared with me months earlier, an acidic constitution *promotes* the spread of cancer, while an alkaline environment *inhibits* cancer.

In contrast to my cancer surgeries and the prescribed radiation and chemotherapy that I refused, the Gerson Therapy left me feeling great! For anyone suffering from cancer or other chronic degenerative disease, I would urge you to consider the

---

85 The Gerson Institute, The Gerson Therapy, www.gerson.org, Friday, September 16, 2011.

Gerson Therapy as an alternative treatment. On the Gerson Institute's website they explain:

> *No treatment works for everyone, every time. Anyone who tells you otherwise is not giving you the facts. We know that when you have been diagnosed with a life-threatening ailment, choosing the best strategy for fighting your illness can be a bewildering task. Everyone claims to have either "the best treatment," "the fastest cure," or "the only therapy that works." In most cases your trusted family physician only has knowledge of conventional treatments, and is either unaware of, or even hostile toward alternative options.*
>
> *No matter how many opinions you receive on how to treat your disease, you are going to make the final decision on what to do, and you must be comfortable with your decision. Choose a treatment that makes the most sense to you.*
>
> *Most therapies–conventional or alternative–treat only the individual symptoms, while ignoring what is ultimately causing the disease. The reason the Gerson Therapy is effective with so many different ailments is because it restores the body's incredible ability to heal itself. Rather than treating only the symptoms of a particular disease, the Gerson Therapy treats the cause of the disease itself. Although we feel the Gerson Therapy is the most comprehensive treatment for disease, we don't claim it will cure everything or everyone.*[86]

---

86 The Gerson Institute, Is the Gerson Therapy Right for You? www.gerson.org, Friday, September 16, 2011.

As far as I know I've been cancer-free since 2008. I thank God for my health and for healing me. I truly credit Him with my recovery. God works in many creative ways, sometimes without benefit of medicine and sometimes in conjunction with it. "Lord my God, I called to you for help, and you healed me." (Psalm 30:2 NIV) I am grateful for the additional years I've enjoyed with my daughters and family!

That being said, it was time to roll up my sleeves and get on to the next exciting project...

Chapter Thirty
— SHOOT FOR THE MOON! —

You're not a loser because things don't turn out, you're a loser if you quit!
Dr. Kenneth Friendly

FOLLOWING THE EVEREST climb and the Gerson Therapy, I came back to Anchorage and continued to care for my daughters. During that time, I helped my nephew, Bart Jr. earn his GED and work through Trend Setters School of Beauty to become a hair stylist. I also worked to pay off all expenses related to the Everest climb and my cancer therapy in Los Angeles.

Sometime later, my friend Patrick McCourt called me from Seattle inviting me to come and work with him on some projects. The former situation with his properties had changed and Patrick decided on a different course of action to restructure debt and return other properties to the bank. Patrick was quickly running out of time due to the downturn in the economy. In Seattle, Patrick introduced me to Mike Cook who owned OSO Lumber, the largest privately held lumber company in the Northwest.

Mike wanted to sell his lumber internationally and asked if I would help him with the marketing. We began marketing his lumber products around the world with a heavy concentration on Asia. Mike and I became good friends and started working out together at a local gym. He asked if I would help him with his son, McKinsey, who was struggling in high school.

McKinsey had a drug addiction (opiates) and had dropped out of school. I started working with him like a drill sergeant and helped him obtain his high school diploma. I also set up a strict regimen for him working out at the gym and eating nutritious food. He was making great progress until he fell off the wagon and succumbed to the drugs again.

Mike and I set up quite a few transactions, but the recession dried up the market. Mike later sold OSO Lumber to a competitor.

Meanwhile, Greg Ellis, my former partner from the World Gym, had staked a 70,000 acre gold claim in Alaska—the Ground Hog and Nika Mines. His mines were directly adjacent to Pebble Mine. The Pebble Gold Mine contains certified mineral reserves of over $500 billion. Greg wanted to redeem himself on account of our broken partnership in the World Gym and some money he owed me. Consequently, he gave me a 25 percent interest in the gold claim, kept 25 percent for himself, and divided up the remaining 50 percent between Patrick McCourt and Patrick's business partner Dusty.

The Pebble mine is touted as containing the largest concentration of gold and copper reserves in the US. The Ground Hog mine shares the border with the Pebble mine's highest concentration of gold. A person could place one foot in the Pebble mine and the other in the Ground Hog mine. I share this to indicate the potential value of our mines. None of my partners nor I planned to actually mine these precious metals. Our goal was to resell the assets for a profit.

In an effort to sell the mines, I placed some ads in the Wall Street Journal seeking potential buyers. We received a couple hundred inquiries that we had to sift through. However, my old friend in London, Leslie Greyling, aggressively committed to make the purchase. His immediate and adamant insistence on buying the mines prevented us from fully considering other offers, but we had taken his eagerness as a good sign. I flew to London and later Patrick and his team joined me to complete the deal. However, finalizing the transaction extended my stay in London, Zurich and other European countries for almost two years as I will detail below.

Due to the delay in closing, we were beginning to feel the financial squeeze. In order to maintain our claim on the two gold mines, we had to pay the State of Alaska $250,000 per year. But because we believed were only days away from closing the sale, we diverted those $250,000 annual fees toward 168 acres of prime development property in Malibu, California rather than send our payment to Alaska. Greg, Pat, Dusty and I all held a financial interest in the Malibu property. This was spectacular ocean-view property. The Malibu property was also under contract to sell—that is, we had identified a buyer before we bought the property.

What I admire most about my brother is his attention to detail and his kindness. Everything he does is for the family, never for himself. He takes care of everybody else.

Also, he's an entrepreneur. He has always had the courage to go for and make something of nothing. He steps out and takes the risk.

He's one of a kind! There's a super strength about him. He's not a big guy, but he has power in everything he does.

For instance, climbing Everest with cancer—most people couldn't climb Everest without cancer. George is an extremely strong person.

– Jerry LaMoureaux, Jr.

Selling the mines was a complex deal that became protracted out and then the buyer failed to perform. I had had some dealings with Greyling in the past and should have known better. I had written a backup deal with a Saudi investor, but he failed to perform as well. This was one of those situations in which we were scrambling to hold things together as the economy crumbled. And this being the height of the recession, the property in Malibu also failed to sell. We had exhausted our cash to retain these assets and maintaining an overseas office was costly.

True, I was striking out, but I had to keep swinging for the fence like Babe Ruth! Babe Ruth was one of the greatest home-run hitters of all-time. But what many do not know is that he was also one of the most prolific strike-out players. He explained once, "Every strike brings me closer to the next home run." So I too kept on swinging. Many multi-millionaires invest in so many different businesses, never knowing which ones would succeed. I couldn't quit. I couldn't give up.

While I was still in London additional international opportunities opened up for me. On May 1, 2009, I assumed the position of Executive Chairman at MinMet, PLC. MinMet, PLC is a global natural resource company with over 6,000 stockholders and substantial mining interests in gold, copper, molybdenum, manganese, iron, uranium, platinum, lead, chrome, nickel, bauxite,

---

My dad, George, is a great man! He has integrity. He's always on to the next goal or target.

Even when things go sour, he keeps on going. He did so with his Karate, the Cartoon Channel—I was with him during The Cartoon Channel.

I asked him after Everest, "What's the next thing you'll conquer?" He said, "Space!"

– Tamra, daughter

zinc, marble, granite, oil, gas, diamonds and other minerals. The company had been in business for over 21 years at the time. It was part of our plan to restructure and rename the company to Monte Cristo Resources.

As Chairman of MinMet, we were under contract to purchase a major portion of the oil and gas reserves in North Dakota. This was before it was widely known and publicized that the Dakotas held these oil and gas reserves. Unfortunately, however, after a short time at the helm of MinMet, I resigned when I discovered what I believed to be less than ethical practices by one of the majority stock holders and I refused to be a part of his schemes.

Their intent was to strip the company of its assets and to "pump and dump" their stock. Through means of deceit, a stock is pumped up (increased in value) through press releases and false claims, after which the stock is sold at its highest price (dumped) on the market. In the end the stock holder is left with nothing.

Having left MinMet, I created Monte Cristo Resources and became Chairman and Chief Executive Officer. As a parallel effort we launched H-3 Energy, Inc. H-3 Energy, Inc. was one of several ventures that we were investigating, not knowing which project we would pursue now, later, or never.

Within H-3 Energy, Inc., I proposed a bold and radical business idea. To some this idea may sound like the fantasy of a science fiction movie, but it *was* viable.

Monte Cristo Resources held substantial interests in a wide variety of precious metals and fuels. We teamed up with Trilliant Exploration Corporation in a "Green Energy Project" under the auspices of newly formed H-3 Energy, Inc. We had named the company "H-3 Energy, Inc." to signify the energy source—Helium-3—that we were going after.

Helium-3 or H-3 is a nuclear isotope that serves as a clean fuel source *without* any radioactive waste. With existing science, Helium-3 produces "cold fusion" and is touted as "A miracle power source of the future." The concept was to use H-3 to power nuclear fusion reactors to produce electricity safely and cleanly.

But here was the challenge with the concept: scientists estimate that there are only a few pounds of H-3 immediately available for extraction on earth. And total reserves amount to less than 15 tons on earth. Where else do we find H-3? On the moon's surface! Scientists believe there are reserves of up to 500 million tons of H-3 on the moon, enough clean fuel to generate all of Earth's energy needs for the next thousand years. It is estimated that two fully loaded Space Shuttles could carry enough H-3 to power the needs for the entire US for a year! Was this a far-fetched idea? Maybe, but why not?

In September, 2009, I drafted the following press release, "Our plan is to be the first commercially viable Space Mission, with the goal to bring clean cold fusion energy to Earth for the future of our children for generations to come." [87] We planned to tap into the cash-starved Russian Federal Space Agency, both because of the frequency of missions the Russians sponsor as well as their lower cost.

I wanted to dovetail the "Missing Children's Telethon" into the lunar mining project by filming and televising the flights

---

87 Monte Cristo Resources press release, September 2009, p. 34.

Shoot for the Moon! | 241

and events, all the while, having ACIM being the beneficiary of the multi-billion viewer event. These flights would've included a Soyuz mission to the International Space Station, an Apollo 13 circumlunar flight around the moon, and a lunar landing. This telethon would have been incorporated into a "Major television event with historic proportions" covering the Denali climb, the Everest climb, and these two space missions.[88]

Following are some details of the two space missions and the Missing Children's Telethon that I proposed:

We will first go into space in 2010 via a placement on the Russian Soyuz Rocket and Space Module to the International Space Station (ISS) and broadcast from there the First Missing Children's Telethon; this will give the Lunar Mission additional credibility and will substantially attract major sponsors for the Lunar event.

In 2012, we will begin our Lunar Space Mission to the far side of the moon by first launching aboard a Soyuz Spacecraft. Then, a subsequent launch will occur of an unmanned Rocket Booster. Our spacecraft will rendezvous with this additional system in low-Earth-orbit. The engagement of the two will provide our Spacecraft with the required propellant to travel to the Moon.

All during the Lunar Missions we will be broadcasting the Missing Children's Telethon, garnering more and more viewers on the way to the "Critical Point" where the Space Module goes around the dark side of the Moon, loses communication and if the trajectory is not exact and on target, the Space Module will be lost

---

88 Letter to Sir Richard Branson from George LaMoureaux, September 21, 2009, p. 2

and never return. Then as we return back home, the world will continue to watch until we parachute land and are interviewed "live."[89]

We had been working with Fox Television Network and negotiated with the Russian Federal Space Agency to implement these plans for televising the Missing Children's Telethon.

In the end, we could not bring the funding together in time to pull off this massive undertaking. I had planned on using my own money for this enterprise by selling copper isotopes. I had contracted to sell the non-radioactive isotopes to Samsung and other advanced technology companies with whom we had contact. Their purchases of billions of dollars' worth of copper isotopes would have more than covered the cost of the space expedition. However, at the time it was not meant to be. I continue to dream and plan how we might conduct this or an equally stunning event to attract viewers to an ACIM telethon.

During the years from 2000 to 2010 I served as a national and global Advisor and Representative of Phoenix Worldwide Industries, Inc. I also helped raise millions of dollars for them. Dr. J. Al Esquivel serves as the President and CEO of Phoenix Worldwide Industries, Inc. Al is a Navy SEAL and was the head of the Phoenix Division, which were the assassin squads. Al holds two nuclear physics doctorates from two universities. One of his doctoral dissertations explained how to build a nuclear reactor the size of a closet. That paper is still classified today. He is an incredibly brilliant man and brother of my good friend Joe Esquivel, who is a financial genius.

---

[89] Lunar 2012 Business Plan, pp. 7 and 9.

Phoenix Worldwide Industries, Inc. manufactures all of the detection platforms for weapons of mass destruction for the federal government. This includes what the Redstone Arsenal uses and trains on. Phoenix Worldwide Industries also builds covert electronic intelligence systems for counter-terrorism and drug interdiction. These detection platforms and high-end intelligence systems are utilized by every branch of our US military and a long list of US national agencies that monitor the security of our nation.

References for Phoenix Worldwide Industries, Inc. include: the Central Intelligence Agency, Department of Homeland Security, United States Army, United States Special Operations Command, United States Joint Communications and Support Element Command, US Department of State, Federal Bureau of Investigation, Drug Enforcement Administration, Bureau of Alcohol, Tobacco and Firearms, Immigration & Naturalization Service, Nuclear Regulatory Commission, Department of Defense, SPAWAR, NAVAIR, Federal Communications Commission (FCC), High Intensity Drug Trafficking Task Forces (HIDTA), Organized Crime Drug Enforcement Task Forces (OCDETF), Food & Drug Administration (FDA), Department of Interior (DOI), Naval Intelligence Service (NIS) & Naval Criminal Investigative Service (NCIS), Department of Justice Federal Bureau of Prisons (FBP), Department of State Diplomatic Security Service (DSS), Redstone Arsenal Missile Test/Telemetry Range, US Panama Canal Administration, General Services Administration (GSA), US Southern Command (USSC), Pax River Test Range, US Naval Warfare (USNW), Office of the Inspector General (OIG), International Law Enforcement and Narcotics Bureau (ILENB), Hazardous Devices School Training Center (FBI), US Department of

State Anti-Terrorism Assistance Program (ATAP), Explosives Incident Countermeasures School (EICS), City, County and State Police Departments throughout the US, various classified projects, and the systems and platforms for the White House for everything the president uses out of the White House's numerous other agencies.

During the two years I spent in London and Zurich working on these ventures, I desperately missed my daughters Ashley, Tamra, my nephews, and the rest of my family. I kept thinking to myself, "One of these days, I'm going to hit one out of the park and be able to take care of everybody." This goal kept me actively pursuing one idea after the next and positioned me for an unbelievable opportunity.

Chapter Thirty-One

— BUYING THE WORLD'S LARGEST LIBRARY OF MOTION PICTURES (MGM)! —

Sow your seed in the morning, and at evening let your hands not be idle, for you do not know which will succeed, whether this or that, or whether both will do equally well.
Ecclesiastes 11:6 NIV

STILL IN LONDON and Zurich, I created another company, the Monte Cristo Entertainment Group, integrating entertainment with advanced technology. The Monte Cristo Entertainment Group has two primary areas of focus: an entertainment industry division and an advanced technology division.

On the entertainment side of the company, I had learned that MGM (Metro Goldwyn Mayer, Inc.) had been purchased some time ago for more than $5 billion. However, MGM now stood on the verge of bankruptcy and I knew I could purchase it at a bargain. But I needed to find an investor with resources of that magnitude. So, I proposed the acquisition of MGM in its entirety to a group of investors in Abu Dhabi and they agreed to move forward on it.

I wrote the MGM Acquisition & Development Executive Summary, dated December 23, 2009. This Summary begins on page two with the following excerpts:

> *The acquisition and development of MGM, its 4,100+ film library and over 10,400 television episodes along with related assets is an opportunity of a lifetime! It is the intent of Monte Cristo Entertainment to acquire in its entirety MGM, its library and all related assets and turn it into the "World's Largest Entertainment Channel."*[90]

In January, 2010, I began correspondence with Moelis & Company, LLC – a global investment bank that was brokering the sale of MGM. At the request of the Abu Dhabi investors, they wished to keep a low profile throughout this transaction, maintaining anonymity.

On April 8, 2010, I received proof of funds in my name for $3 billion committed to me from the Abu Dhabi investors for the acquisition of MGM. They awarded me this proof of funds based on my resume, background checks, performance with The Cartoon Channel, other successful ventures, and billions of dollars of proven assets that I held in my name and/or my companies' names at the time.

When I created Monte Cristo Entertainment, I put together a substantial board of directors and advisors who were well known in the entertainment industry. At this time, my board and I flew to L.A. and met with Moelis & Company and made our concrete proposal to purchase MGM for $2.5 billion, keeping $500 million in reserve. I gave substantial time, money and energy to this project as it was my plan to build the world's largest television channel using MGM's library.

I had obtained a tremendous amount of assets, but needed cash flow to maintain them. The cash flow from the

---

90 George LaMoureaux, Monte Cristo Entertainment, Inc. Acquisition Confidential Overview of Metro-Goldwyn-Mayer, Inc., December 23, 2009, p. 2.

acquisition of MGM and launching of the Entertainment Channel would have helped me hold onto those assets while I was in the process of liquidating them. I use assets that are doing nothing to acquire loans for business ventures, as long as the business ventures support themselves and the assets simultaneously. This is the art of arbitrage: securing financial assets and turning them around to sell them for a profit.

The purpose of acquiring MGM Studios was to leverage it to become the "World's Largest Entertainment Channel—bringing laughter and adventure your way 24 hours a day!" I saw this as my second chance to fulfill what I had started when I created The Cartoon Channel and Toon TV.

The promotional points for marketing the World's Largest Entertainment Channel lauded: *The world's best movies; 24 hours a day; streamed instantly free.* (The channel would've been paid for by advertising.) To launch and host this entertainment channel, we set up the website: www.mgmfree.com.

An excerpt from a letter I wrote to a soldier I know who needed encouragement:

July 16, 2010

Dear —,

Actually, there were a few things happening in my life. 2007 was a rough year. Within a few months my brother died, my wife left me and then I found out I had cancer and endured five cancer surgeries. But then God blessed me with the ability to climb Mt. Everest. While on the mountain I received word that my sister had been murdered.

I came back from climbing Everest and had my sixth cancer surgery. A few weeks later, my nephew died. Then, over the next few months I lost three of my uncles.

In December of 2009, I lost everything I owned, including my house and car.

Even so, I feel blessed as I have made my reservations with Christ. Make sure you make yours too.

God bless you and yours. I am praying for you as you go back to war.

– George LaMoureaux

The entertainment industry can be an extremely lucrative investment. For instance, the overall budget for producing the *Lord of the Rings* series was $281 million. But that series had grossed $2.9 billion by 2010. The revenue projections for this entertainment channel over a five year period exceeded $56 billion!

The acquisition of MGM, Inc. was moving along nicely until the seller decided it wanted to draw the Abu Dhabi investors into the limelight over the purchase. But the Abu Dhabi investors insisted on anonymity. They told me, "George, we will back you all the way, but keep our name out of it." With more direct pressure from the seller on this issue, the Abu Dhabi investors backed out of the deal withdrawing their offer to fund the purchase. You can imagine my disappointment!

Steve Jobs once said, "Being the richest man in the cemetery doesn't matter to me. Going to bed at night saying we've done something wonderful, that's what matters to me.[91] Money in and of itself means nothing to me. I want to do something wonderful, beneficial, and God-glorifying, so an integral element of all of these ventures sought to provide funds for A Child Is Missing.

At the same time I was pursuing the MGM purchase I was making headway on the advanced technical side of the Monte Cristo Entertainment Group in another direction. I had acquired a 90 percent ownership interest in a substantial reserve of copper powder isotopes Cu63 and Cu65. These non-radioactive nuclear isotopes are highly sought after in very specialized fields of aerospace, high-end electronics, nuclear medicine, and laboratory use.

---

91   BrainyQuote.com, Steve Jobs.

Copper isotopes are traded in grams. The price of a gram of copper isotopes depends greatly on the market, particle size and purity. We happened to own those of the highest grade. At the time, we were offering these high-end copper isotopes at a market price of $4,353.90 per gram. I had a million grams (1,000 kg) for sale amounting to approximately $4.3 billion.[92] This was one of the reasons I received the $3 billion dollar proof of funds from the Abu Dhabi investors for the purchase of MGM.

Due to the possibility of scams or improper storage and handling, provenance is of utmost importance in the sale and purchase of these isotopes. Our copper isotopes were held at a facilities designed for this purpose in Germany, Switzerland and Poland. They issue a "Safekeeping Receipt" with full details of the assets, certifying their authenticity for a potential buyer or lender.

Storage is tricky and therefore important. The isotopes have to be sealed in airtight packaging with argon gas. Any exposure to oxygen ruins them. There is a flourishing black market since the price is so high and the powder is easily concealed and transported.[93]

While in London and Zurich, multi-millionaire Paul Ryan, was brokering the sale of copper isotopes on my behalf in South Korea, China and Japan. We had contracted the sale of these isotopes to Samsung and other major technology companies during this period. These were multi-billion dollar contracts.

---

92 George LaMoureaux, Letter from Monte Cristo Entertainment Group dated May 31, 2010.
93 George LaMoureaux, CSM Review, supplement to letter dated May 31, 2010, p. 3.

At one point, Paul Ryan threw a huge weekend event in my honor at his country home north of London. Paul wanted to show his appreciation to me because he was one of the beneficiaries in a sale of my copper isotopes. Paul treated me like royalty, chauffeuring me around in his Maserati for the weekend. But I was especially impressed with Jenny Bae, one of the celebrity entertainers Paul had engaged for the event. Jenny is a world-renowned violinist who has performed internationally for over a decade. And to think that Paul brought Jenny in specifically to play for me at this event!

At the weekend gala, Jenny and I became attracted to each other. Later, on my return to London, Jenny called me and wanted to get together, so we began seeing each other on purely a friendship basis. In the end I had to admit that I didn't have the cash flow to run with her globetrotting crowd. I needed to concentrate on my family in Anchorage while she was inviting me to travel the world with her, including meeting her parents in South Korea. I had the assets, but not the cash flow. I don't think she was interested in me for the money, as she left her multi-billionaire fiancé without concern for money.

On occasion Jenny and I still write to each other. As it happens, she was also the former fiancée of the heir to the Samsung fortune, although my connections to Samsung had nothing to do with my relationship with her.

Over the years, I've negotiated copper isotope contracts with Samsung and other Asian, European and US based countries, and more recently with Pakistan. However, these high grade copper isotopes can be used in the production of nuclear weapons. Therefore, due to the instability of the region, I terminated these negotiations and other proposed transactions with the government of Pakistan in which I was representing Phoenix Worldwide Industries in the sales of advanced intelligence systems.

I continue to broker the sale of these copper isotopes today and the final story has yet to be told on the remaining reserves of these isotopes.

Additionally, Paul Ryan had introduced me to an attorney who connected me with the royal family in the country of Oman. This attorney issued to me $500 million in gold from the royal family to be used in trading programs in Switzerland. We created companies in Switzerland to enable these transactions. However, we could not effectively transfer the ownership of the gold because of the complexities associated with the foreign banks and their various requirements for the trading program. So we returned all of the assets.

Late in 2010, after spending more than two years in London and Zurich, I longed to be home with my daughters and near my family. Even though I still had transactions pending, I became overwhelmed with the desire to come back to Anchorage for my family. I missed my daughters Ashley and Tamra deeply and my father's health had

> One of the most remarkable characteristics of my dad is that no matter where he has been, or where I was, he always called me every single day. He even called me daily on his satellite phone from Everest!
>
> When I was a little girl, living with my mom in California, dad called one night while I was lying on my bed. His soothing voice soon put me to sleep.
>
> The next thing I knew, Mom came into my room accompanied by two police officers. When I fell asleep on the phone, Dad called the police and had them check on me to ensure I was alright!
>
> – Ashley LaMoureaux, daughter

deteriorated so that he needed caring for. Consequently, it was my father's health that triggered my immediate return to the US.

However, I didn't realize just how bad my father was until I returned to Anchorage. There I dropped most everything I was working on and simply spent time with my dad and cared for his needs. He was on dialysis and had other substantial requirements, so I elected to serve him in our family home as his live-in caregiver. I really felt this was the honorable thing to do.

We should learn first of all to put our religion into practice by caring for our own family and so repaying our parents and grandparents, for this is pleasing to God. – 1 Timothy 5:4

Especially because Jerry, my step-dad, had stepped into our big family and had taken me on as his son, loving me and caring for me all those years, this was the least that I could do for him.

Chapter Thirty-Two
— I FELT LIKE JOB! —

*As you know, we count as blessed those who have persevered.
You have heard of Job's perseverance and have seen what the
Lord finally brought about.
The Lord is full of compassion and mercy.*
James 5:11, NIV

WITH MY PLANS to return to Alaska, I had hoped to somehow complete my outstanding international business transactions. But shortly before leaving London, Norman Bailey, my Swiss banker died. I had leaned on him heavily in setting up international companies and conducting international transactions.

Then, after I returned to Anchorage, my attorney, James Loughran, died in London. Without my banker and my lawyer, I thought, "Well, at least I still have my accountant." But then my accountant/CPA, Terry Petruska, passed away too! I was beginning to feel like Job in the Bible!

While in London, I had sold many of my assets in the US and used those funds to extend my stay in Europe and put money into the ventures I was pursuing with my business partners. This rendered me cash-poor now that I was back in Anchorage.

At that time, I was supposed to fly to Korea and meet with Samsung to complete the sale of copper isotopes. But being without my banker, attorney and CPA complicated matters significantly and because I could not physically be present in Korea to complete the transaction, it fell through. I was at a loss as to how to meet my international obligations long distance.

I consoled myself that at least I was home with Ashley, Tamra and my parents. As the oldest son, I really felt responsible to take care of my dad. During our childhood, our parents sacrifice so much for us that we either take for granted or aren't even aware of. In view of all my parents had done for me, taking care of them in their old age was the right and honorable thing for me to do. Yet I still had to make a living.

In Anchorage, while taking care of my father, I was working long-distance to close some bond transactions that I had taken on. Additionally, in order to provide income for my family, I helped my brother Max build another branch of his insurance business.

Then, due to some previous business transactions I had managed, I acquired 84 percent interest in $850 million in emeralds and $568 million in artwork. The owner retained 16 percent. The purpose of taking on these assets was to leverage them in order to acquire financing and to place them into trading programs, a percentage of which would have gone to support A Child is Missing, much the same as I had done with previous enterprises.

But after meeting with numerous individuals about this trading project, I felt uncomfortable with the amount of risk involved. I held the fiduciary responsibility for these assets and needed to act in a way that protected everyone's interests. So I returned 100 percent of the assets to their owners.

Add all of these business deals to my responsibilities caring for my dad and spending time with Ashley and I had a very full schedule. In the middle of all this an old friend, Keith, came to me for help.

Keith and I had known each other since my days owning The Ritz. At that time we were dating sisters—he with a young woman named Shelly and I with her sister Paula. Years later, it was Keith who had suggested I sell real estate for a while, which I had done successfully. I had also helped him with the purchase of a gym.

When I returned to Anchorage in 2011, my friend was operating that gym. This was one of the gyms that I helped build years before. Then when it sold, I brokered the deal for the second owners. Finally, I had helped Keith with the deal when he and his family purchased this gym.

As things stood now, Keith and his family were the current owners of the gym, but were being forced to relocate it. He needed to find a new property and wanted to build a brand new facility. Keith kept calling me for assistance and I helped him here and there and told him what to do. But I had too much going on my life to devote any more time to his project.

Then in mid-August Keith invited me out to dinner one night. Over our meal he pled with me to help him with the lease of the new property and with the financing and construction of his new facility. I knew this would be a full-time endeavor, because I had already built or been involved in the creation of five gyms prior to this on my own: Body Tech Gym, the Powerhouse Gym, and the World Gyms in Lake Stevens, Monroe and Everett.

I had an intimate understanding of what would be involved from now until opening and knew it would demand my full attention. But I also saw that Keith needed my help to obtain financing.

In spite of the heavy demands this project would make on my time, I thought, "Well, perhaps this is a good opportunity to help a friend down on his luck and earn some cash right now until my other deals come through."

So, I jumped in and helped Keith and his family and began pulling the financing together, coming up with a new name, creating the logo, many facets of the design of the gym, the equipment, and its layout. Keith and I called on a number of contractors, some of whom I had worked with in the past. I always made an effort to select the best contractors *and* negotiate good rates.

There were many days that I went without sleep for up to two days to stay on top of things and bring the construction to a timely completion. When we opened the new gym in October that year—just three months later, it was a really big deal locally. The mayor of Anchorage was there to cut the ribbon on this state-of-the-art, gorgeous facility.

However, during this project, my dad's health began to worsen. Then one day he fell out of bed trying to get up and broke his hip. He had a short hospital stay and then they moved him to an extended care facility as his needs exceeded what we could provide for him at home. Even so, every day he was there, I went to see him and spend time with him. In early January, 2012, Dad passed away and went home to be with the Lord.

Following my dad's death, I moved out of my parent's home and into my own place. My mom had grown accustomed to having me around, but my international calls at all hours of the day and night were an intrusion on my family. We all needed our privacy.

I was receiving some income from another business deal from the past, but would have to figure out a way to earn more cash while I continued to work on the other business opportunities that I had in the works. Also, my dad's retirement was not passed on to my mom, so my brother Max and I increased our efforts to care for our mom financially.

At this season of my life, I found myself in a real financial bind. I never intended to take a job, but under the circumstances with my dad's death and my mom's loss of income, I needed cash. Even though I had business opportunities in the works overseas, I could not provide the level of support that my international investors and business partners required. My international ventures would have demanded that I jump on a plane at a moment's notice and perhaps be gone for months. I didn't want to do that to Ashley and my family again. So, I resorted to something I hadn't done since I was 16.

Chapter Thirty-Three
— THE TALLEST CROSS —

The majority of men meet with failure because of their lack of persistence in creating new plans to take the place of those which fail.

Napoleon Hill

IN VIEW OF my circumstances and need for immediate cash flow, I reluctantly considered an option to which I was not accustomed. A friend of mine shared with me that he was making ten-to-fourteen thousand dollars a month selling cars at a major auto dealership in Anchorage.

I had sold more than a hundred of my personal cars privately and as a teen had sold cars professionally for a short period of time. Some years earlier, I had proposed to buy Worthington Ford from my friend Cal Worthington. Cal and I had done cross promotions with each other when I owned Church's Fried Chicken. But back then, instead of pursuing the auto business, I had gone the route of the nightclubs—a huge mistake!

Taking this job at this major auto dealership was very difficult for me. I hadn't worked as an *employee* since I was 16. I had always been the *employer*. I recoiled at the thought of being on someone else's payroll after all these years.

But I needed a regular income to cover my expenses and selling automobiles would allow me to be home a lot and spend time with my daughters Ashley and Tamra. However, I went to work at this auto dealership kicking and screaming. I still had hundreds of millions of dollars of assets in my name, but I lacked cash flow.

Considering the positions I've held—right or wrong—I was embarrassed to sell cars. By comparison, I felt like I was going to work at McDonalds! By that comment I don't mean anything derogatory toward the dealership I work for or the industry in general. But this was very difficult for me—stifling!

The auto group that I work for is a family owned business whose owners treat me like one of them. On the job, I get paid to shake hands, kiss babies and make people happy. This dealership is the largest of its kind in Alaska and they're a good company to work for. As of this writing, I've been a Sales and Leasing Professional with them for about three years. I and another over-achieving co-worker of mine compete for first and second place for highest sales volume each month. We've made a game of it.

I make a good six-figure income, so the job affords me the opportunity to maintain a lifestyle I enjoy. But most importantly, I rarely travel and I'm home with Ashley every day and see Tamra frequently. Since working in auto sales, I've considered the purchase of three different auto dealerships that have been offered to me. But I've decided that owning a dealership would distract me from closing the other deals I'm already working on.

In addition to these other business ventures I'm seeking to close, I've worked on two other projects—one not-for-profit the other for profit. The non-profit proposal is to erect the nation's largest cross on the campus of the Anchorage Baptist Temple

(ABT). There are two proposals actually. One is for a 230-foot, free-standing steel cross that would be the largest in the nation. The other proposal would be for a 250-foot, free-standing steel cross that would be the largest of its kind in the world.

I have nothing to gain financially from this project. Also, it would be very easy to spiritualize this project involving a huge cross and to argue its merits and drawbacks purely from a spiritual point of view. Obviously, because the cross symbolizes Christianity and what Christ has done for us, the cross *does* carry spiritual significance. But I simply contend that setting up the tallest cross in America or the world on ABT's campus is a strategic and much needed marketing tool for drawing people to the church and bringing them to Christ.

As an entrepreneur and marketing professional I've made the following observations. First, I'm told that every five years, sixty percent of the population of Anchorage turns over. So we're constantly dealing with an influx of new military personnel and those employed by the large oil companies, the hospitals, and other businesses. This doesn't even account for the 1.8 million tourists that Alaska is privileged to host each year.

I'm convinced that the nation's tallest cross on ABT's campus would attract people who would not have otherwise known there is a church there. And as people are drawn to ABT via the cross, those who come will put their trust in Christ and the church will grow numerically, financially, and in every other respect.

I saw this enormous cross as a way to boldly proclaim our allegiance to Christ and draw others to Him. May the reader feel my passion behind this project!

Second, I believe we're losing our moral bearings as a nation. We're going the route of Sodom and Gomorrah. While elsewhere in the country factions misapply the Constitution to tear down crosses in public places, perhaps this cross could serve as a symbol that there are still many who follow Jesus Christ and that our country was founded on Christian principles and values. President Ronald Reagan said, "If we ever forget that we are 'One nation under God,' then we will be a nation gone under."

Dr. Adrian Rodgers remarks about the giant crosses that Bellevue Baptist Church in Memphis, Tennessee, erected some years ago:

The crosses are the talk of the town. I cannot tell you what a blessing it is to drive up on Sunday night and see those crosses illuminated against the sky. I am in the mood to worship when I get there. They say—come in and hear about Jesus.[94]

The proposed cross would be a beautiful structure, fabricated out of steel with a white finish. I had contracted with The Headrick Company, an organization that specializes in designing and fabricating such structures for churches across America.

In my presentation to ABT I explained:

ABT's majestic cross plays a major role in projecting a positive image to our members and the community. After my many months and intense conversations and work with the Headrick Company, I believe that Headrick is dedicated to providing us with a quality majestic cross that effectively conveys an image of

---

94 Dr. Adrian Rodgers as quoted in The Hedrick Company brochure, p. 2.

strength, security and stability to Anchorage, Alaska, America and throughout the world.[95]

Intent on seeing this cross built, I've been in talks with ABT and Pastor Jerry Prevo now for over three years. We already have FAA approval and preliminary approval from the municipality to erect the cross at ABT.

The 49-page, spiral bound proposal contains: demographic research for marketing purposes; a detailed appeal for erecting the cross; proposed plans for funding the project; artist renditions of the cross on ABT's property; architectural drawings of the cross and its base panels; Headrick's prospectus; and numerous testimonials, references, and pictures from Headrick's satisfied clients.

Preliminary designs of the proposed cross include four beautifully sculptured panels that form the base of the cross. These sculpted panels would depict key biblical scenes or events. At the base of the cross would be a large courtyard on which people could congregate.

On March 30, 2012, Channel 2, KTUU News delivered the following news brief:

The Anchorage Baptist Temple is exploring the option of adding a new landmark to the skies over East Anchorage by the end of next year: a 230-foot-tall cross built on its property that could be the world's tallest, according to a request for an aeronautical study filed Feb. 24 with the Federal Aviation Administration.

---

95 George LaMoureaux, Presentation of the Cross, p. 4.

> *The monopole cross, which would be manufactured and installed by Laurel, Miss. construction firm Headrick Inc. according to data the temple filed with the FAA, is slated to include a 55-foot-long cross beam 192 feet above the ground.*[96]

Although flattering, Channel 2 News was mistaken about our cross vying for the rank as the tallest one in the world. The tallest cross in the world is in the Valley of the Fallen in Spain. That stone cross reaches 500 feet into the sky and can be seen 30 miles away!

Over these past three-plus years I have been in communication with the builder of the 190-foot Cross in Groom, Texas, who tells me he has 1,000 to 2,000 visitors a day who stop to look at the Cross. They say that the Groom, Texas Cross can be seen from as far as 20 miles away!

As should be expected, when the news leaked out that we were even considering raising a 230- or 250-foot cross at ABT, we began receiving both negative and positive feedback from individuals in the community. However, I firmly believe that the benefits of building the tallest cross in America at ABT will substantially outweigh the criticism and challenges we will face.

I believe that this cross will bring worldwide attention to the ministry of ABT and will draw many people to Christ. Setting up this cross and taking advantage of what it represents could also bring revival to the church. Furthermore, through the influence of the cross, I believe that Pastor Prevo's sermons

---

96 Christine Kim and Chris Klint, "Anchorage Baptist Temple Plans to Build 230-Foot-Tall Cross," Channel 2 News, March 30, 2012, http://articles.ktuu.com/2012-03-30/cross_31268384.

will gain local, national and international attention resulting in furthering the Gospel of Jesus Christ.

Obviously, finances would pose a major challenge for any church in building America's or the world's tallest free-standing steel cross. However, I have been in direct communication with Stealth Concealment, a company that specializes in concealing broadcast and communication towers. If we can get the Municipality of Anchorage to approve ABT as the location for the tallest cross in America, perhaps we could fund its construction in part by incorporating concealed broadcast antennas into the cross.

Meanwhile, I have acquired all the blueprints for other similar crosses that The Headrick Company has designed and constructed. Along with these were 468 pages of engineering calculations and 100 pages of blueprints!

We have yet to see what the outcome will be and whether Anchorage will be home to the nation's (or world's) tallest free-standing steel cross. But I continue to pour my energies into this project believing in its huge benefits.

Chapter Thirty-Four

— FROM CROSS TO CROSSHAIRS —

If you don't have a sword, sell your cloak and buy one.
Luke 22:36 NIV

AT THIS POINT in my story, I'd like to make a monumental transition from a *cross* to *crosshairs*. My brother Max and I are always on the lookout for additional businesses to launch. The fact that we liked guns led us to open Lethal Weapon Guns, a store specializing in the largest selection of tactical weapons in the state at the time.

We had operated a brick and mortar storefront for nearly a year. However, in 2013, we decided to redirect the business entirely to the internet, because neither Max nor I had the time or desire to manage a physical store. More recently, Max and I sold Lethal Weapon Guns for a profit.

Prior to closing the brick and mortar storefront, a disturbing incident took place in the parking lot of Lethal Weapon Guns. A few months ago, I was driving across town in my

---

In one of my businesses, I hired a consultant to work with our team. He asked us to consider the following question: If your ship was sinking and you could only take five people with you, who would they be?

George is without question the first guy I'd have in my lifeboat. I'd choose him because of his faith, tenacity, and creativity. He's very calm under crisis.

*– Patrick McCourt,
Entrepreneur and friend*

Mercedes when I received an urgent call from Max. He was at Lethal Weapon Guns and said that there was a man outside going berserk. Max had already called the police. Minutes later I arrived in the parking lot to witness a bazaar scene.

A very large Black man, about six-foot-five and 250 pounds, in his mid-to-late twenties was jumping up and down on the hood and roof of a pickup truck belonging to a friend of ours. C.C. , the owner of the truck, is also a large Black man of perhaps 300 pounds. C.C. was trying to talk the other man off his truck. This act of violence was unprovoked and the two men did not know each other.

We could not dissuade the perpetrator from destroying C.C.'s truck, so I told C.C. to drive off. But as C.C. tried to drive away, the man would not get off his truck. C.C. didn't want to hurt him, so he stopped, parked, got out and grabbed the perpetrator's leg and tried to pull him off the truck. C.C. jumped on top of him to hold him down, but the man threw C.C. off like a rag doll. At this, the perpetrator became even wilder, pounding C.C. in the face, jumping down on top of him beating him up.

This guy had a crazed, demonic look in his eyes that told me he had to be on some very nasty drugs. He also wielded almost super-human strength displayed by the damage he had done to C.C.'s truck and the way he was knocking big C.C. around. All of this happened very quickly.

I had gotten out of my car and was standing there wearing my Rolex watch and a thousand-dollar suit that I didn't want to soil. But C.C. was getting pounded and started calling for help. So I tore my watch off, threw it in the car and ran over and subdued the perpetrator with a martial arts hold. I didn't hurt him, but I had control over him so that he couldn't move.

C.C. was able to get up while I continued to hold this guy down. I spoke very calmly to the perpetrator telling him that I wasn't going to hurt him, but to calm down. Meanwhile, we were still waiting for the police to arrive, which took some 20 minutes. Max held the perpetrator's legs down, because he was trying to kick me in the back with his heels.

When law enforcement did show up, an officer came over and put a handcuff on one of the perpetrator's wrists. I told the cop that he'd better cuff both his hands and that he was extremely strong and crazed.

The officer was cocky and told me he had control of the situation, so I released my hold on the man and jumped away. Immediately, the police had a fight on their hands! It took four officers to get the other cuff on him and they had to zip-tie his legs together as well. The police suspected that this guy was on either bath salts or meth—both very bad drugs.

Chapter Thirty-Five
— THINKING BIG! —

You have to think anyway, so why not think big?
Donald Trump

As you can probably guess, I resonate very strongly with that quote above from Donald Trump. I also particularly like his statement, "Anyone who thinks my story is anywhere near over is sadly mistaken!" My track record over the past three years bears that out.

Now some might think, "George is a guy who dreams big and has accomplished some pretty grand things in his prime, but now he's just selling cars." To set the record straight, let's just say that I sell cars right now because it offers me the freedom to pursue much, much greater goals. Foremost among those goals is being home to spend time with my two daughters.

Additionally, I'd like to highlight two of several projects in which I've invested considerable time and money in the past three years—while I was "just selling cars." Again, the auto group that I work for is a family owned business whose owners treat me like one of them. On the job, I get paid to shake hands, kiss babies and make people happy. This dealership is the largest of its kind in Alaska and they're a good company to work for.

The Pebble Mine

Remember, in the months following my ascent of Everest, I acquired 25 percent interest in a 70,000 acre gold claim in Alaska—the Ground Hog and Nika Mines. I had mentioned that these mines were directly adjacent to Pebble Mine, the granddaddy of gold mines in Alaska.

The Pebble Gold Mine lies north of Iliamna Lake in southwestern Alaska. This area is a vast, unpopulated expanse of marsh and shrub. The Pebble Gold Mine contains certified mineral reserves of over $500 billion—$120 billion of that in gold as well as copper and molybdenum.

To be more specific, 1200 test holes drilled in the area have demonstrated that the Pebble deposit represents 6.44 billion tons of measured and indicated resources containing 57 billion lbs. copper, 70 million oz. gold, 3.4 billion lbs. molybdenum and 344 million oz. silver; and 4.46 billion tons of inferred resources, containing 24.5 billion lbs. copper, 37 million oz. gold, 2.2 billion lbs. molybdenum and 170 million oz. silver.[97] This makes the Pebble Project "one of the world's greatest stores of mineral wealth."[98]

Well, a couple of years ago, I made a bid to purchase the Pebble Gold Mine for $8 billion. In order to facilitate this deal, I set up The Pebble Limited Partnership in Alaska. And I identified an interested foreign investor who was willing and able to front the $8 billion.

---

97   Northern Dynasty Minerals Ltd. http://www.northerndynastyminerals.com/i/pdf/ndm/NDM_FactSheet.pdf.
98   Northern Dynasty Minerals Ltd. http://www.northerndynastyminerals.com/ndm/Home.asp.

Thinking Big! | 273

The goal of the Pebble project was to develop and operate the gold mine in an efficient manner, such that it would very handsomely reward the three major financial partners in the project:

- The foreign investor (whom I will leave unnamed) would own 50%
- Northern Dynasty (along with several smaller investors) would own 40%
- Monte Cristo Resources (I, George LaMoureaux) would own 10%

As CEO of Monte Cristo Resources and liaison of the project, I would receive a draw in the amount of 10 percent of the total project funding ($800 million) and 10 percent of the total project equity. Monte Cristo Resources would also manage and oversee all work, labor, services, materials, utility, transportation and other acts necessary as performed and expended by the mine's operating partner to complete and manage the power plant in a workmanlike and efficient manner.[99]

This $8 billion was to be invested over a one year period in three phases:

1. **Phase I Funding - $2.5 billion USD:** Phase I funding was to be provided upon obtainment of formal contract with Northern Dynasty Minerals Ltd. and other owners in which the foreign investor obtains desired level of equity in the project.

2. **Phase II Funding - $ 3.0 billion USD:** Phase II funding needs would be requested and provided six

---
99 "Pebble Gold Mine Project Agreement," May 2014, George LaMoureaux.

months in advance of demonstrated project commitments, to ensure efficient project continuity.

3. **Phase III Funding - $2.5 billion USD:** Phase II funding needs would be requested and provided six months in advance of demonstrated project commitments, to facilitate efficient project continuity.

Monte Cristo Resources would ensure the completion and effective management of the project on behalf of the investors, generating substantial returns for the funder, and to a lesser extent, Monte Cristo Resources.[100]

Now enter President Obama and the EPA and you can guess where the project went from there—nowhere! The Washington Post reported in February 2015, "As early as this spring, the Environmental Protection Agency is expected to invoke a rarely used legal authority to bar a Canadian company, Northern Dynasty Minerals Ltd., from beginning work on its proposed Pebble Mine, citing risks to salmon and to Alaska's pristine Bristol Bay, 150 miles downstream."[101]

Although we could demonstrate that we could operate Pebble in an environmentally responsible way, there were just too many political hurdles preventing its progress at the time. Perhaps in the future…

The Trans-Alaska Gas Pipeline

Since the summer of 1977, oil has been flowing the 800 miles from Prudhoe Bay on the North Slope of Alaska

---

[100] "Pebble Gold Mine Project Agreement," May 2014, George LaMoureaux.
[101] Joby Warrick, "Pebble Mine Debate in Alaska: EPA Becomes Target by Planning for Rare 'Veto'," The Washington Post, February 15, 2015, http://www.washingtonpost.com/national/health-science/internal-memos-spur-accusations-of-bias-as-epa-moves-to-block-gold-mine/2015/02/15/3ff101c0-b2ba-11e4-854b-a38d13486ba1_story.html.

southward to the harbor in Valdez, Alaska. Construction began on that project on March 27, 1975 at a cost of $8 billion.[102]

It took several years of debates, hearings, injunctions, and the creation of a nine-volume, 3,500-page environmental impact statement to get the project moving.[103,104]

About half of the pipeline is elevated to prevent the warm oil from melting the permafrost and causing environmental damage. Approximately 950,000 barrels of crude oil flow through the pipeline each day.

Then, in March 2014, the governor of Alaska, Sean Parnell announced a proposal to build an 800-mile natural gas pipeline from the North Slope of Alaska to Nikiski on the Kenai Peninsula.[105] The natural gas reserves along the Beaufort Sea on the North Slope of Alaska represent one of the largest untapped reserves in the world.[106]

The estimated cost of this project was projected between $45 and $65 billion.[107] The question is who will fund this massive project?

Solving questions like that is one of the things I live for! Consequently, I identified both local and foreign investors who were anxious to get behind this project. In August of

---

102 Michael Ray, "Trans-Alaska Pipeline," Encyclopedia Britannica, May 16, 2014, http://www.britannica.com/topic/Trans-Alaska-Pipeline.

103 Wikipedia, "Trans-Alaska Pipeline System," nd, https://en.wikipedia.org/wiki/Trans-Alaska_Pipeline_System.

104 Jack Rafuse, "Alaska Gas Project Dwarfs Proposed Keystone XL Pipeline," The Washington Examiner, March 1, 2014, http://www.washingtonexaminer.com/alaska-gas-project-dwarfs-proposed-keystone-xl-pipeline/article/2544897.

105 Lisa Demer, "Alaska Senate Passes Measure to Move Gas Pipeline Forward," Alaska Dispatch News, March 18, 2014, http://www.adn.com/article/20140318/alaska-senate-passes-measure-move-gas-pipeline-forward.

106 Jack Rafuse.

107 Lisa Demer.

2014, I prepared the Trans-Alaska Gas Pipeline Memorandum of Understanding and created the Trans-Alaska Pipeline Corporation, "for the development, construction and operation of the 800 mile, forty two (42) inch, Trans-Alaska Gas Pipeline."[108] This pipeline would deliver up to 4.5 billion cubic feet of natural gas daily.

Once again, my corporation Monte Cristo Resources was to provide liaison services on behalf of our investors, on-site project leadership, management, project services and program status feedback.

Because this model works well, I proposed a three-phase funding process for the Trans-Alaska Gas Pipeline:[109]

1. **Phase I Funding - $2.5 billion USD:** Phase I funding would be provided upon obtainment of Formal Contract and Final Agreement with the Trans-Alaska Gas Pipeline Corporation (TBF).

2. **Phase II Funding - $ 7.5 billion USD:** Phase II funding would be provided upon obtainment of formal contract with other owners, which will include the State of Alaska and may also include, Exxon, Conoco Phillips, ARCO and/or others by mutual agreement. Phase II funding needs will be requested by the Trans-Alaska Gas Pipeline Corporation (TBF) and provided by our foreign investors seven (7) months in advance of demonstrated project progress, to ensure efficient project continuity.

---

108 George LaMoureaux, "Trans-Alaska Gas Pipeline Memorandum of Understanding," August 2014.
109 George LaMoureaux.

3. **Phase III Funding - $55.0 billion USD***: Phase III funding needs would be requested by the Trans-Alaska Gas Pipeline Corporation (TBF) and provided by our foreign investors twelve (12) months in advance of demonstrated project progress, to ensure efficient project continuity. (*Funding requirements may be less than the total of $65 billion US dollars projected for the overall cost as other participants/owners may provide funding.)

In association with the Trans-Alaska Gas Pipeline Corporation, Monte Cristo Resources would manage and oversee all work, labor, services, materials, utility, transportation and other acts necessary as performed and expended to complete and manage the Trans-Alaska Gas Pipeline Corporation in a workmanlike and efficient manner.[110]

Monte Cristo Resources was to receive a payment amounting to 10 percent of the total investment ($6.5 billion). Equity ownership was to be divided among three entities as follows:

- Our foreign investor group – 65% ownership
- Monte Cristo Resources (George LaMoureaux) – 10% ownership
- The State of Alaska (possibly to include Exxon, Conoco Phillips, ARCO and others) – 25%

I put provisions in place to alter those percentages based on level of investment should the oil companies choose to shoulder a portion of the development.

---

110 George LaMoureaux.

During this process I met with State of Alaska Governor Sean Parnell and Congressman Don Young several times and others regarding the funding of the Trans Alaska Gas Pipeline.

All documents for funding were signed by the foreign investors including phased funding and bank-to-bank wire transfer agreements. Everything was ready to go to launch the Trans Alaska Gas Pipeline, but due to a major political shift in the country of our investor group and the substantial drop in global oil and gas prices, the investors and their banks decided not to fund the project at this time.

My investors/partners take the position more similarly to a banker than a participant in the limelight of acquisitions, administration and operations. These investors are more concerned about the collateral, mitigation of any possible loss, and their return on investment (ROI) than anything else.

Also, all federal, state and local clearance of good funds received by me and my companies always include adherence to the Patriot Act's anti-terrorist and anti-money laundering rules and regulations along with the Dodd-Frank banking and securities conformances.

We hope to revisit this opportunity should the oil and gas prices rise again.

These and nearly all my projects I have always run in an extremely confidential manner. I have seen many a project fail even before it gets off the ground due to big mouths flapping! However, due to the current status with these two projects and the fact that I'm writing this book, I chose to disclose them here.

Recently, in addition to the above, my good friend Patrick McCourt offered me an ownership position in his Fracking Supply Company in the Dakotas. He flew me there attempting to entice me to accept his gracious proposal, but I turned it down. Accepting this offer and others I've received from overseas would take me away from my daughters and family—something I'm no longer willing to do.

But by writing these things, I hope thereby to inspire and empower others to "Dream big and dare to fail," as my friend and supporter Colonel Norman Vaughan has always said.

Chapter Thirty Six
— NOT A SPECTATOR —

*Greater love has no one than this:
to lay down one's life for one's friends.*
John 15:13 NIV

I DO EVERYTHING I can to stay away from trouble. But I don't sit on my hands. I'm not a spectator. I engage. Similarly, I took action in a different kind of incident during the writing of this book. "There are three types of people in this world: those who make things happen, those who watch things happen and those who wonder what happened." I aspire to be the first type of person who make things happen.

On August 24, 2013, I was in the process of selling a vehicle to a client when we both heard the engine stall on a single-engine airplane that had just taken off from Merrill Field a couple blocks from our dealership. We both looked up and to our horror saw the small plane careen into the ground on the airstrip.

The following account appeared in the Alaska Dispatch two days later in which I was interviewed by a reporter:

> *"Bart, my customer, pointed to the sky and said, 'Look at that plane, its engine just cut out'," LaMoureaux said Monday. I looked up to see a flash of its white wings as it plunged straight down."*

A huge thud followed, and LaMoureaux and his customer ran for a nearby truck.

"We drove about 500 feet to the fence (surrounding Merrill Field) and I hopped out, and before I knew it, I was over the fence, running towards the plane," LaMoureaux said.

"The plane was crumpled, with the engine smashed in, and tail broken. I could see a small fire starting near the front," LaMoureaux said. That's when he saw Lilly, the plane's pilot, unmoving and tangled in the wreckage.

"I tried to push on the wings to open up the cockpit, but no matter how much I tried, I couldn't get him free," he said.

As LaMoureaux jostled the wreckage, a woman fell from the back of the plane, onto the pilot.

"I could see she was strapped in, but neither she nor the pilot were moving or making any noise," LaMoureaux said.

The fire, now working its way into the mangled cockpit, caught LaMoureaux's attention again, just as two people were running toward him.

"Get some fire extinguishers. I need fire extinguishers," he yelled.

As Anchorage Police began to show up, LaMoureaux said he had just doused the flames, with three fire extinguishers people had brought to him, taken from nearby parked aircraft.

> "They began CPR on both people, but I could tell they didn't make it," LaMoureaux said.
>
> "I just wish I could have done more. I did everything I could think of, but it wasn't enough," he said.
>
> On Monday, LaMoureax was back at work. A small bandage on his right hand covered some cuts he got when jumping the airport fence, but the cuts were superficial. The events of a few days ago made a deeper scar on the 57 year old.

> A few days after the plane crash at Merrill Field, my dad showed me where it had happened.
>
> I looked up at the tall fence with barb wire at the top and asked him, "How did you get over that barb wire?" He said he didn't even think about it at the time, but that's how he cut his hand.
>
> – Ashley LaMoureaux, daughter

> "(Jessi's) family called me to thank me for trying to help their daughter. The only thing I could say to them to ease burden, was that I don't think she suffered. I think they (Nelsen and Lilly) died, almost instantly," LaMoureax said.[111]

I didn't think twice about scaling a nine-foot barbed wire fence trying to save those two young people. My only thought at the moment was to stop the fire before the plane exploded and prevent them from burning to death. I wish things had turned out differently for them.

About a year later, I received a phone call from the police department. At first I thought, "What have I done now?" But they explained that they wanted to recognize me for my act of

---

111 Sean Doogan, "Grieving Young Couple Victims – Alaska's Most Recent Fatal Plane Crash," Alaska Dispatch, August 26, 2013.

heroism. I received an official letter from the Anchorage Police Chief, Mark T. Mew and an invitation to an event at which a plaque was awarded me.

The Police Chief's letter read as follows:

Dear Mr. LaMoureaux,

On August 24, 2013, you performed an exceptionally courageous act in the saving of a human life that displayed conspicuous initiative and outstanding attention to duty. The Anchorage Police Department extends its recognition and appreciation and presents to you the commendation of the

CITIZEN'S CERTIFICATE OF VALOR

Officers Jonathan Butler, Matthew Barth, Eric Christian and civilian George LaMoureaux responded to a catastrophic crash of a Cessna 150 aircraft at Merrill Field. The aircraft had impacted the ground nose first, crushing the engine compartment and bending the fuselage in half. They immediately recognized that there were two people on board, still trapped in the wreckage, and that smoke was rapidly filling the passenger compartment from a fire in the engine compartment. They all ran to the burning aircraft without regard to their own safety knowing the aircraft could explode at any moment. Mr. LaMoureaux worked the fire extinguisher and the officers heroically worked to extricate the two occupants from the wreckage, dragging them away from the immediate danger areas and began performing lifesaving efforts until the arrival of the AFD. The officers endured the choking smoke, expanding fire and fire extinguishing agents to free the occupants. Although neither of the occupants survived, the efforts of these officers and civilian demonstrate courage and compassion despite the danger to themselves.

Sincerely,

Mark T. Mew
Chief of Police

Some months following the Police Chief's letter, I also received notice of a resolution put forth by the Anchorage Municipal Assembly "recognizing citizen George LaMoureaux for the Anchorage Police Department Partners in Public Safety Citizens Award for Valor. This resolution states:

> *Whereas, on August 24, 2013, Citizen George LaMoureaux joined several APD officers to respond to a catastrophic crash of a Cessna 150 aircraft at Merrill Field; and*
>
> *Whereas, the officers and Citizen George LaMoureaux immediately recognized that there were two people onboard, still trapped in the wreckage, and that smoke was rapidly filling the passenger compartment from a fire in the engine compartment; and*
>
> *Whereas, the offers and Citizen George LaMoureaux ran to the burning aircraft without regard to their own safety, knowing the aircraft could explode at any moment; and*
>
> *Whereas Citizen George LaMoureaux worked the fire extinguisher while the officers worked to extricate the two occupants from the wreckage; and*

Whereas, although neither of the occupants survived, the efforts of these officers and Civilian George LaMoureaux demonstrated courage and compassion despite danger to themselves.

Now, therefore, the Anchorage Assembly hereby:

Recognizes and applauds Citizen George LaMoureaux with the Anchorage Police Department Partners in Public Safety Citizens Award for Valor.

Passed and approved by the Anchorage Municipal Assembly this 7th day of October, 2014.

(Signed by both the Chair and Municipal Clerk.)

Fast forward to December 4th, 2015, another day in the life of George LaMoureaux. I was driving home at about 9:45pm, when the pickup truck in front of me went out of control on the icy road, hit a guard rail, flipped, careened into the ditch and landed on the driver's side. I witnessed the accident and was the first one on the scene. I quickly pulled over, got out of my truck and ran down to the crashed vehicle.

Inside, I found a young, teenage girl in hysterics. Remarkably, she appeared unharmed except for some minor cuts on her hands and arms. I was anxious to get her out of the smoldering truck before it caught fire and into the warmth of my vehicle until the paramedics could arrive.

As I was pulling her from her vehicle another car stopped to help and I directed them to call 911. Once inside my vehicle, the girl continued to cry hysterically and I did what I could to calm her down. Because her purse and phone were still in her vehicle, I asked her for her parents' phone number so I could call them and let them know that she was okay.

When the police and fire department arrived, I turned the young girl over to their care and gave them a report of what had happened. When I arrived home, the young girl's parents contacted me and thanked me for saving their daughter's life. I thank God that He allowed me to be on the scene and take action!

There's yet another way in which I seek to take action, which the following illustrates. Back in 2012, I had written Donald Trump urging him to run for President. On June 16, 2015, Donald Trump announced that he would run for President of the United States. People laughed at him!

I had previously supported Ronald Reagan and was personally working with Nancy Reagan in support of "Just Say No International." I had contributed hundreds of thousands of dollars in advertising spots and committed millions in perpetuity in support of the War on Drugs during the days of the Cartoon Channel. Later, I campaigned for Mitt Romney and then for John McCain and Sarah Palin in Washington, Nevada and Alaska. I've always supported the conservative Republican agenda. Furthermore, I campaigned for Senator Dan Sullivan, Amy Demboski, a Republican running for Anchorage Mayor, and Congressman Don Young. I also campaigned for Sean Parnell on his Alaska Governor re-election campaign and supported many others.

So, in April 2016, I attended the Alaska Republican Convention in Fairbanks as a State Delegate. This came on the heels of an interview with Alan Cobb who was with the Donald J. Trump for President National Campaign. At the Republican Convention in Fairbanks, Jim Crawford was named the Chairman of the Donald J. Trump for President Campaign for Alaska. Jim had been the Chairman for the Ronald Reagan for President Campaign in 1980 and 1984. Jim has a list of many other accomplishments including the Finance Chairman of the George H.W. Bush Campaign in 1988 and the Chairman of the Bob Dole for President Campaign in 1992. While I was at the Alaska Statewide Republican Convention, I was named the Statewide Volunteer Coordinator for the Donald J. Trump for President Campaign. My appointment was confirmed by Jim Crawford and former Alaska State Senator Jerry Ward, who went on to be the Statewide Director of the Campaign in Alaska.

The acceptance of this position, the Statewide Volunteer Coordinator for the Trump Campaign in Alaska, was a heavy responsibility. I was already carrying a full load in my professional career and with my other businesses. Even so, I felt compelled to support Donald Trump because I believe that our country was going in the wrong direction. In fact, this brought a quote to mind that I seek to live by. As the great parliamentarian, Edmund Burke said, "The only thing necessary for the triumph of evil is for good men to do nothing."

I felt overwhelmingly compelled to jump in with both feet and help Donald Trump become our President and not to merely sit on the sidelines.

At that time, we faced a substantial challenge in Alaska, because neither of our Senators, Lisa Murkowski nor Senator Dan Sullivan, were supporting Trump for President.

Additionally, our Congressman Don Young was not openly supporting Trump.

Furthermore, the Alaska Federation of Natives (AFN) had endorsed Hillary Clinton, our Governor was not a Republican, nor was our liberal Anchorage Mayor a supporter of Donald Trump for President.

Then, in July of that year, I attended the Republican National Convention in Cleveland, Ohio. There I was privileged to meet with Newt Gingrich, Dr. Ben Carson, former Arkansas Governor Mike Huckabee, New Jersey Governor Chris Christie, former US Ambassador John Bolton, and Corey Lewandowski. I told Mr. Lewandowski that we should focus on what he said early in the Trump Campaign and "Let Trump be Trump." Lewandowski named his book by the same name.

In addition, I met with Reince Priebus the Chairman of the Republican National Committee (RNC), who spent some time with me and was excited about receiving a copy of my book.

There were so many people that I met with in my support of Trump's Campaign for President. Following is a short list of some of those people: Sean Hannity, Judge Jeanine Pirro, Jesse Watters, Sheriff David Clarke, my old friend Don King, Wayne Allen Root, Larry Elder, Ted Koppel and another longtime friend Geraldo Rivera, Bret Baier, Megyn Kelly and Charles Krauthammer, Chris Wallace, Lisa Kennedy, and briefly and in passing Donald Trump Jr., Eric Trump and Ivanka Trump to whom I gave my Climb for America's Children Commemorative Coins. All were very gracious to me.

I worked day and night to help move Trump's Campaign forward here in Alaska and on a national level. And I was privileged to meet with Mitt Romney and Ted Cruz when they came to help promote Trump's campaign in Alaska.

In fact, early on I constantly told my longtime friend Dr. Jerry Prevo that he should support Donald Trump for President. Dr. Prevo is the Pastor of the Anchorage Baptist Temple in Anchorage, Alaska. His church is one of the largest and most powerful churches in Alaska. Dr. Prevo is also the Chairman of Liberty University with over 110,000 students in attendance. Dr. Prevo had met with every President from Ronald Reagan to George W. Bush, but by choice did not meet with President Obama due to the direction he was taking our country. My Mom always said, "Pastor Prevo is the most courageous man in Alaska!"

I gave a copy of Donald Trump's then newly released book, "Crippled America," to Dr. Prevo's wife. And throughout the early part of the campaign, I asked Dr. Prevo numerous times who he was going to vote for. He repeatedly told me that he wasn't sure yet. Even so, he eventually told me that his wife Carol said, "I don't care who you vote for Jerry, but I am voting for Donald Trump!"

Of course, that made me happy! This news also encouraged me as I've always felt Dr. Prevo listened to the leading of God and would make the right choice.

I would continue to give him pertinent updates in-person and via text messages about the campaign and what Donald Trump was doing. Finally one day, Dr. Prevo said that he was going to vote for Donald Trump for President, "because we just may have a friend in the White House!"

Then something miraculous happened! Sometime later, Jerry Falwell Jr, President of Liberty University had Donald Trump speak at Liberty. At that time, Jerry Falwell announced that he was supporting and formally endorsing Donald Trump for President. Dr. Falwell said, "Because we just may have a friend in the White House!"

When I heard about Jerry Falwell's endorsement of Trump for President, I knew that Pastor Prevo had something to do with it. Dr. Falwell had used the same words as Dr. Prevo had in his endorsement of Trump. I asked Dr. Prevo about his involvement in Jerry Falwell's endorsement of Donald Trump and he admitted that he had talked to him about it. Of course he had! And Jerry Falwell Jr. was using Dr. Prevo's words.

Then Dallas megachurch pastor, Robert Jeffress, was interviewed and said that he was supporting and endorsing Donald Trump for President because, "We just might have a friend in the White House!" And again, Dr. Prevo's words echoed in my mind...

Then, Dr. Prevo's good friend, Franklin Graham, who owns a lodge here in Alaska, came out in support of Donald Trump for President.

Later in the campaign, I had put in a phone call to Donald Trump through the National Campaign headquarters to suggest that Franklin Graham and Dr. Prevo say the Opening Prayer at the Cleveland, Ohio Republican National Convention. The response and request came back explaining that they would like to see Franklin Graham and Dr. Prevo give the closing prayer. And although that was not to be, Franklin Graham did say the prayer at the Presidential Inauguration in Washington, DC!

Simultaneous with these efforts on a national level, I was an editor and contributing ghostwriter for social media websites supporting Trump with over 250 million views a month. All of this was done without any fanfare or publicity directed toward me. Furthermore, I launched one of the most impactful websites of the campaign and remained anonymous in my association with its funding. There was a small team of people who are true patriots that worked tirelessly with me who, like me, don't want the credit associated with our efforts other than to see a greater and better America for generations to come!

In Alaska, we had a Spartan team of Trump supporters that I brought on board. To name a few, they included my two daughters, Ashley and Tamra, the rest of my family and friends, Mike Robbins, Kenneth Bottcher, Carla Hendrix, Skeeter Hansen, Al Ferretti, and Randy Comer. They are all committed and die-hard Trump supporters! There were many other volunteers, but the aforementioned were some of the key supporters and hard workers in the campaign.

We had placed more than 7,000 campaign signs across Alaska. In fact, we put out so many signs that a new ordinance was put into place fining the campaign for signage placed outside of the now legally imposed parameters. In a single night we would put out hundreds of signs, which were taken down the next day by Hillary Clinton supporters who did not want to see Trump become President. Every day was a challenge, including driving steel spikes into the frozen ground to put up the signs.

I met with influential individuals and promoted Mr. Trump any way I could. This included the Alaska State Fair with over 293,424 people in attendance. At the fair, we had a booth where we passed out thousands of bumper stickers and campaign signs. We also distributed "Make America Great Again" campaign buttons, T-shirts and hats.

My daughters and I set up every major Trump-for-President-Campaign event and rallied campaign volunteers to wave signs at every major intersection.

In the campaign process, I was interviewed by most every major news network in Cleveland, Ohio and in Alaska. Among them was the Anchorage Daily News (ADN). They interviewed me on election night, November 8th, 2016 at the Trump Headquarters in downtown Anchorage. The ADN article said in part:

> *Trump Alaska campaign workers gathered for a victory party at Flattop Pizza in Downtown Anchorage, where they're celebrating.*
>
> *George LaMoureaux, Trump's Alaska Volunteer Coordinator, said, "Trump has the right plan to rebuild the country and 'Make America Great Again.' Hillary Clinton," he said, "is corrupt, a serial liar and an evil woman."*
>
> *LaMoureaux said he supported Trump's views on trade and building a wall along the Mexican border, which he said would protect America from drugs coming into this country.*
>
> *"If it's part of God's plan, he will be President," he said.*

The rest is history and Donald J. Trump became the 45th President of the United States. I was delighted to receive an invitation and attend his inauguration in Washington, DC on December 20, 2017.

I can absolutely and unequivocally say that if it had not been for the Evangelical vote and a part of God's plan, Donald Trump would not be President!

Given the condition of our country, I knew I had to do something. As it says in James 4:17, "Anyone then who knows the good he ought to do and doesn't do it, sins." Consequently, I believe it would have been a sin of omission for me had I not gotten involved in helping Donald Trump get elected to help turn our country around.

And for anyone who thinks that I'm blind in my devotion to Donald Trump, please know that I read 20 or more news stories every day from the AP to the Wall Street Journal and I am fully informed as to what's happening.

I truly feel that I was blessed to play a small part in helping Donald Trump become our President!

Chapter Thirty-Seven
— BE A FINISHER! —

Though the righteous fall seven times, they rise again.
Proverbs 24:16 NIV

MOST BIOGRAPHIES ARE written posthumously. The unique (and tough) thing about an autobiography is knowing when to stop writing and to publish, because life goes on! Much has happened in my life during the writing of this book and no doubt much will continue to happen.

For example, just days before publishing, I received a phone call from my good friend and sensei, the Grand Master Sensei Robert V. Alejandre, who could hardly contain his excitement. It seems that he had hardly slept the night before in anticipation of the news he was about to share. In his sly, Grand Master way, he informed me that he had promoted me to "Fifth Degree Master Black Belt," the highest rank within his system! He cited this as a result of more than 30 years dedication as a Black Belt. He confessed that he should have awarded me this rank five years ago.

His announcement took me completely by surprise, similar to the call I had received from the Anchorage Police Department regarding the Citizens Award for Valor.

I would like to deflect this great honor toward God that He might receive the glory. All credit goes to God for healing and strengthening me and for bringing me to where I am. Had it not been for God, I wouldn't be here to accept this great honor. And this brings the Scripture to mind, "For I can do all things through Him (Christ) who strengthens me." —Philippians 4:13

Meanwhile, aside from the everyday excitement in my life, I've held myself back from pursuing numerous large, fast-paced business deals for the past few years. Instead, I'm trying to maintain some stability with my daughters Ashley and Tamra and being here for my mom and other family members. Terri, Ashley and Tamra's mother, and I are still very close. Each Sunday for the past three years, I've enjoyed taking Ashley and my mom to the morning and evening services at Anchorage Baptist Temple where Dr. Jerry Prevo serves as pastor.

I have high regard for Pastor Prevo. He is absolutely sold out to Jesus Christ. He stands for what the Bible says and doesn't allow himself to be swayed by political correctness or the winds of culture. He preaches straight up. My mom

---

I met George here at ChangePoint. He's been a volunteer greeter and in security. He has a real heart for God and a passion for life. Christ keeps him centered on that.

George is a defender of life—a registered lethal weapon. He sees a threat and he steps into action like no one else I've ever seen. He has a heart to be a servant. Ask him and he'll do it.

One time someone came out of the men's room saying that a man in a stall had a gun. George went in and confronted the situation. He doesn't fear for his own safety.

He dresses impeccably. He's a real gentleman. He's consistent, trustworthy and cares for people. George is also fascinating—a mystery.

He has a huge heart and a colorful past! He's not ashamed to share Christ with others and tell them what Christ has done in his life.

– Karl Privoznik, Connections Pastor

adored him and always said, "Pastor Prevo is the most courageous man in Alaska."

During the writing of this book, my dear mother, Mia LaMoureaux, passed away Sunday, September 8, 2013 and went home to be with the Lord. Mom died of natural causes in her home gym, wearing her sweat suit and weight lifting gloves. She was lying on her Bowflex® Bench, dumbbells by her side. Mom was "hard as nails and sweet and kind as an angel." We held Mom's memorial service at Anchorage Baptist Temple. She was a wonderful mother and we all miss her greatly.

> "It doesn't matter what you say, Dad. It matters how you make others feel. That's what they remember the most."
>
> – Ashley LaMoureaux, daughter

In addition to attending Anchorage Baptist Temple Sunday evenings, I'm often there for the Sunday morning services both to worship and to volunteer. I serve as incognito security. I enjoy serving others.

Being aware of others' problems has also given me greater perspective on my own troubles and helps me focus on someone other than myself. Of the heartaches, losses and challenges I've faced, I would rank the top five in the following order beginning with the most difficult:

1. The loss of so many family members
2. Being away from my daughters Ashley and Tamra and my family for two years while I was in London and Zurich
3. Fighting cancer
4. Climbing McKinley and Everest

5. The loss of my wife, finances and having to start over again

When I look over my list and think of the tragedies in others' lives, the thing that stands out in my mind is how little time we have on this earth. The realization of how fragile and brief our lives are leads me to ask questions like, "What legacy will I leave? How will I influence and impact others for good? When I stand before Christ, will He say to me, 'Well done, good and faithful servant'"?

Anyone who doesn't wrestle with these questions is engaging in some method of self-medicating activity to avoid reality. We all leave a mark here. The question is what kind of mark will we leave?

To a degree, my past makes me what I am today. But I am not shackled to my past. Christ gave me a do-over, a second chance. God has also slowed me down to write this book.

In the Introduction, I mentioned that I needed a noble purpose for writing this book. Why would anyone want to read about my life? I told you that I came up with not just one, but three noble reasons for writing.

First, my sincere hope is that by reading about the losses, hardships, and failures I have experienced and overcome that you would be inspired and motivated to persevere. Never quit! Keep on swinging for the fence!

Also I realize that my experiences may not speak as clearly to you as they do to others. So pick up biographies of other men and women who have been beaten down, endured much and continue to get up and keep on trying. Remember Ken

Friendly's words, "You're not a loser because things don't turn out, you're a loser if you quit!"

Read about others like: George Washington, Abraham Lincoln, William Wilberforce, President Ronald and Nancy Reagan, George H. Bush and George W. Bush, Henry Ford, Preston Tucker, Eric Liddell, Dietrich Bonhoeffer, Jackie Robinson, Howard Hughes, Thomas Edison, Dale Carnegie, Napoleon Hill, Martin Luther King, Albert Einstein, President Harry Truman, Louis Zamperini, Helen Keller, Barbara Bush, Mother Teresa, and others. Allow yourself to be mentored by them. (I've listed a number of books in the Bibliography at the back of this book.)

Second, as I have given myself to advance the non-profit organization A Child Is Missing, I encourage you to find a lofty cause for which to fight and live. Leave a noble legacy for others. There was a time in my life in which I wanted to be known as the owner of The Ritz nightclub, or the founder of The Cartoon Channel, or the buyer of MGM Studios (which would have launched the world's largest entertainment channel), or even the sponsor and benefactor of ACIM. But the real joy comes not from being known for doing great things, but simply serving others.

Today, I continue to work on a number of projects while I maintain an income. These include a $5 billion dollar hydro-electric project and other business ventures not listed.

Finally, my third noble purpose for writing this book overshadows the first two. After reading my story, you've seen where I came from and where my life without Christ was leading me. You also read about the profound change that Christ has made in my life as I follow Him and trust Him. And He continues to transform me.

I urge you not to go your own way. Our culture is obsessed with every person "finding their own truth." By accepting such thinking we've abandoned logic, morality and reality. There are absolute truths in this world. To think otherwise is to deny reality. Therefore, I challenge you to explore the claims of Christ. Read the New Testament. You may think you know what it says, but do you really?

Pick up a book like The Case for Christ by Lee Strobel and read it with an open mind. Strobel, a graduate from Yale Law School was a former atheist and legal editor for the Chicago Tribune. Yet when he set out to disprove the "myth" of Christianity once and for all, he found himself confronted by the truth and kneeling at the feet of Jesus.

Many are afraid of what they will lose if they come to Christ. That's the serpent's age old lie that he hisses in our ear. Don't believe it! No adventure, no amount of pleasures, or money, or power, or anything else this world has to offer even comes close to the wonder and joy of knowing Christ.

Our worth is not found in what we've accomplished, or what we own. Our worth comes from who we are before God. When we come to Him and trust Him through Christ, He makes us His child. Think of that! I'm a child of the Most High God! What could be grander?

The Apostle Paul wrote, "I once thought these things were valuable, but now I consider them worthless because of what Christ has done. Yes, everything else is worthless when compared with the infinite value of knowing Christ Jesus my Lord." (Philippians 3:7-8 NLT)

God blessed me with a few more short years on this earth. Who knows? Maybe it was so I could write this book, tell my story and lead others into a relationship with Him. I pray that you're one of those.

Meanwhile, I'm not done here yet. I keep swinging for the fence. How about you?

Let me leave you with two great encouragements. The first is a favorite quote of mine from President Theodore Roosevelt and the second is Psalm 138, a Psalm of David.

> *It is not the critic who counts; not the man who points out how the strong man stumbles, or where the doer of deeds could have done them better. The credit belongs to the man who is actually in the arena, whose face is marred by dust and sweat and blood; who strives valiantly; who errs, who comes short again and again, because there is no effort without error and shortcoming; but who does actually strive to do the deeds; who knows great enthusiasms, the great devotions; who spends himself in a worthy cause; who at the best knows in the end the triumph of high achievement, and who at the worst, if he fails, at least fails while daring greatly, so that his place shall never be with those cold and timid souls who neither know victory nor defeat.*
> – President Theodore Roosevelt

Anyway

*People are often unreasonable,
illogical, and self-centered;
Forgive them anyway.*

*If you are kind,
people may accuse you of selfish, ulterior motives;
Be kind anyway.*

*If you are successful you will win some
false friends and true enemies;
Succeed anyway.*

*If you are honest and frank,
they may cheat you;
Be forthright anyway.*

*If you find serenity and happiness
they may be jealous;
Be joyful anyway.*

*The good you do today
they often forget tomorrow;
Do good anyway.
Give the world the best you have and
it will never be enough;
Give your best anyway.*

*You see in the final anaylsis,
it is between you and God;
It was never between you and them
Anyway.*

– Mother Theresa

Psalm 138 (NIV), A Psalm of David

I will praise you, Lord, with all my heart;
before the "gods" I will sing your praise.
² I will bow down toward your holy temple
and will praise your name
for your unfailing love and your faithfulness,
for you have so exalted your solemn decree
that it surpasses your fame.
³ When I called, you answered me;
you greatly emboldened me.

⁴ May all the kings of the earth praise you, Lord,
when they hear what you have decreed.
⁵ May they sing of the ways of the Lord,
for the glory of the Lord is great.

⁶ Though the Lord is exalted, he looks kindly on the lowly;
though lofty, he sees them from afar.
⁷ Though I walk in the midst of trouble,
you preserve my life.
You stretch out your hand against the anger of my foes;
with your right hand you save me.
⁸ The Lord will vindicate me;
your love, Lord, endures forever—
do not abandon the works of your hands. (NIV)

APPENDIX

Never Quit!
Condensed Version

Dr. Kenneth Friendly

with Rob Fischer

Foreword by
George LaMoureaux

Scripture taken from the New King James Version®. Copyright © 1982
by Thomas Nelson, Inc. Used by permission. All rights reserved.
Unless otherwise noted, all Scripture references taken from the NKJV.

THE HOLY BIBLE, NEW INTERNATIONAL VERSION®, NIV® Copyright © 1973,
1978, 1984, 2011
by Biblica, Inc.® Used by permission. All rights reserved worldwide.

The ESV® Bible (The Holy Bible, English Standard Version®) copyright © 2001 by
Crossway,
a publishing ministry of Good News Publishers. ESV® Text Edition: 2011.
The ESV® text has been reproduced in cooperation with and by permission of Good News
Publishers.
Unauthorized reproduction of this publication is prohibited. All rights reserved.

Scripture quotations taken from the Amplified® Bible,
Copyright © 1954, 1958, 1962, 1964, 1965, 1987 by The Lockman Foundation
Used by permission. (www.Lockman.org)

We secured special permission from the author to publish this condensed
version of his book, *Never Quit!* in order to include it George LaMoureaux's
autobiography, *Everest — A Triumph in Adversity*
A True Story of Faith in the Face of Extreme Adversity.

TABLE OF CONTENTS

Foreword 309

Chapter One— Never Quit 311

Chapter Two— The Pitfalls of Quitting 315

Chapter Three— Quitting Is a Character Flaw 319

Chapter Four— Get Past the Urge to Quit 325

Chapter Five— Find What God Wants You to Do and
Stick to It! 329

Chapter Six— There Is Victory for Those
Who Persevere 335

Chapter Seven— We Have the World Overcomer
Residing Inside of Us 339

Chapter Eight— Cultivate a Finisher Mindset 347

Chapter Nine— The Unstoppable Spirit 355

Chapter Ten— Set Your Will to Win! 363

Chapter Eleven— The Motivation
of Pain & Pleasure — 371

FOREWORD

ABOUT EIGHT YEARS ago, within just a few months, my brother and two other close family members died; I lost my wife; experienced financial ruin; and was fighting for my life through stage-four cancer. As I was recovering from my fifth cancer surgery and trying to climb out of the emotional and physical hole in which I found myself, I happened to turn on the radio one day and heard the message you're about to read.

I'm not exaggerating when I tell you that Dr. Ken Friendly's message served as the foundation for my personal recovery from cancer and all the other personal tragedies I had so recently experienced. At the time, this material was available only in audio format.

After hearing Dr. Ken Friendly's four-part series on the radio, I drove to his church, Lighthouse Christian Fellowship here in Anchorage, and purchased that message series, *Never Quit!* Over the next three months I must have listened to it a hundred times! Through his message of hope and perseverance, I emerged from grief, despair and hopelessness to a place of health—emotional, physical and spiritual.

In fact, just three months after my fifth cancer surgery, I climbed and summited Mt. Everest. To a great extent, I credit Dr. Ken Friendly's *Never Quit* message with my dramatic recovery and this monumental personal accomplishment.

During my adult life I have been through countless motivational courses and programs including: the Success and

Motivation Institute, Dale Carnegie, Napoleon Hill, Tony Robbins, and so many others that I can't even recall them all today. *Never Quit* stands above them all in my opinion, as pure genius and inspired by God's Word.

A few months ago, I contacted Dr. Ken Friendly about getting his message series into print to which he graciously agreed. I felt compelled to publish his book to inspire others to reach their goals—whatever they are and to never quit!

— George LaMoureaux

Chapter One
— NEVER QUIT! —

And let us not grow weary while doing good, for in due season we shall reap if we do not lose heart.
Galatians 6:9

WE HAVE BECOME a generation of quitters! We quit high school. We quit college. We quit our marriages. We quit our jobs. We quit our weight loss program. We quit our exercise routines. We've become a nation of quitters!

I used to be a chronic quitter. In my earlier days, I would quit at the drop of a hat! I would look for any excuse to quit. But in recent years I've learned to be a finisher. I've learned to endure and to persevere.

As you read this book, I pray that my words will encourage you. Perhaps you have been a quitter like I was. But you too can be a finisher. I exhort you to never, ever quit! Keep on doing what is right.

Quitting is the number one killer of dreams, visions, and callings. There is nothing that can shut us down like quitting. Think about it, quitting is so easy to do. Throwing in the towel isn't difficult. Unfortunately, so many people quit.

Forgive me if I sound a little blunt or mean, but I don't have any other way to say this: *A person who habitually quits is a loser.* You don't lose until you quit.

Yeah, I know, not everything is going to go right. Not everything is going to work out, but that's okay. That's the way life is. You are not a loser because things don't work out, or because you make mistakes. *You're a loser if you quit!*

You don't drown by falling in the water; you drown by not getting out! A lot of people fall. But the Bible tells us, "For though the righteous fall seven times, they rise again." (Proverbs 24:16 NIV) Even the righteous fall, but they get back up. They don't quit!

So, I want to encourage you not to quit. I don't care what the issue is in your life. Don't quit! If you have one more breath, there is still hope. If something is taking a little longer than you thought it was going to take, don't sweat it – it's going to be alright. But if you quit, it's over.

God admonishes us in the Psalms to look to Him to supply our food in *due season* (Psalm 145:15). There is a season for everything and every season has a beginning and an end. What you are going through right now is *temporary*. And if you don't quit, right on the other side of your perseverance and your endurance is the blessing of God.

Napoleon Hill wrote a book called *Think and Grow Rich*. The basis of his book is that he interviewed about 500 famous and very successful people. He wanted to find a common denominator that made them successful.

Hill interviewed the top successful people who were alive at that time—including Henry Ford and many others. When Hill had concluded his tour of interviews, he identified one characteristic that was common in the lives of all these top achievers.

Every one of these successful people was at the point of *quitting* when their breakthrough came. *But, they didn't quit!*

When you are pursuing your dream, all hell will break loose! All hell will break loose in your mind, in your family, and in your finances. You can count on the fact that all hell will break loose.

Quitting doesn't discriminate with age or gender. It doesn't discriminate with political affiliation. It doesn't care. The devil wants to get folks to quit. But if you can hold your course and be a finisher, on the other side of your diligent endurance God has a life that is beyond description.

Where are you in your life and circumstances right now? Do you feel like quitting? If you're honest with yourself, would you like to quit right now, because it's easy? Oh, it is. Quitting is the easiest thing to do.

In Luke 9:62 we read, "But Jesus said to him, 'No one, having put his hand to the plow, and looking back, is fit for the kingdom of God.'" Take note of what Jesus is saying here. If our mindset toward some task is, "I'm going to *try* this." If our mindset is, "I'm going to *give this a shot*," then we'll never make it. We'll end up quitting.

Our mindset has to be, "This is what I'm going to do! I'm going to hold my course, come hell or high water!" If we never, ever give in to the urge to quit, *our persistence will cause our opposition to whimper at our feet.* We must resolve to say, "I'm going to see this thing through, I don't care what happens! I know God asked me to do this."

I'm so tired of folks saying, "God told me I'm going to be this." Or, "God called me to do this." Or, "God gave me this assignment." And then six months or a year later, they are nowhere to be found.

Every godly resolve will be challenged! Every godly resolve is going to be challenged, because the devil knows if he can derail us, he can get us to quit. This is just a test of what God can trust us with. And what God has for you and me is much bigger than what we have right now. But we can never get to the *bigger* and *better* if we can't handle the *now*.

I want to challenge us from Luke 14:28, "For which of you, intending to build a tower, does not sit down first and count the cost, whether he has *enough* to finish it."

Jesus said, before you start, think about *finishing*. While you are here (at the beginning) you've got to have your mind *there* (at the finish line). Don't even start if you're not thinking about finishing.

I remember when we started this church. We started planning to launch this church about two years before we actually established the church. We had the finished church in mind before we started.

I remember telling my wife that as we set out to plant this church, we would have to count the cost. I had to find out if my wife and I were together on this. A husband and wife must work together as a team. Obviously and thankfully she has been with me in all of this.

I think the reason a lot of people quit is because they never actually sat down and counted the cost of finishing. "What's this thing going to cost me? What's it going to cost me?" Jesus said, "Before you start, sit down and count the cost to make sure you can finish."

Chapter Two
— THE PITFALLS OF QUITTING —

*Let us hold fast the confession of our hope without wavering,
for He who promised is faithful.*
Hebrews 10:23

EVERY TIME WE quit we lose something. When we quit, we tip our hand to the devil and he knows how to get us to quit the next time. When we quit, we lose ground.

Every time we start and finish something we are building a resume. And, every time we quit something we're building a resume. One resume is for God, and the other resume is for the devil.

While we are pursuing a goal, God is finding out what He can *trust* us with and the devil is finding out what he can *stop* us with. So, we can't take quitting lightly. Because every time we quit we're building that resume and every time we quit we're building that script. And so the next time, the devil knows how to hit the replay button.

Every time we quit we build *compromise* into our hearts. And when we do that, we are setting ourselves up to quit again. And then quitting can become a *habit*. And before we know it, we're quitting on our wife or husband. We quit on our boss. We'll even quit on God.

When we quit it causes our confidence to diminish. And the danger of losing our confidence is that we come to the point where we don't even try. Because we never complete anything and we never stay with anything, our confidence diminishes. Then, we lose confidence to do even the things that we *can* do and we lose hope.

When we quit we establish a comfort zone. When quitting becomes totally acceptable to us, quitting becomes our comfort zone. Our comfort zone is like a regulator that keeps us at a certain place in life. And if we try to go outside of our comfort zone, bells and whistles go off inside us. And we think, "I can't do this!" And we retreat back to our comfort zone.

That's why we see some people experience a degree of success and then their success starts to bother them. And they sabotage their success in order to get back to their comfort zone.

Once our comfort zone is set, it's difficult to change, but we can *stretch* it and *expand* it. Do you know how we begin to stretch our comfort zone? By not quitting!

Let's shift gears here and look at *how we can persevere.* In John 4:31-32 we read, "In the meantime His disciples urged Him, saying, 'Rabbi, eat.'" Jesus was hungry and needed to eat something. "But He said to them, 'I have food to eat of which you do not know.'" Then in verse 33 we read, "Therefore the disciples said to one another, 'Has anyone brought Him *anything* to eat?'"

Jesus responded, "'I have food to eat of which you do not know. My food is to do the will of Him who sent Me, *and* to finish His work." What Jesus was getting at is this. His "food" was that which energized Him, nourished Him and kept Him going. Jesus not only resolved to *do* the will of God, he had a resolve to *finish*. What nourished and energized Jesus was not only *doing* the Father's work, but also *finishing* it.

A lot of times we are gung ho about starting. We can get all fired up about doing the will of God! But what about finishing it? That's why Scripture tells us, "For we have become partakers of Christ if we hold the beginning of our confidence steadfast to the end." (Hebrews 3:14) Remember the confidence in Christ you started with? Hold fast to your confidence in Christ steadfast to the end.

In Hebrews 10:23 we read, "Let us hold fast the confession of *our* hope without wavering, for He who promised *is* faithful." Outlast the urge to quit, resolve to persevere.

Now it may take me a lifetime, but I'm finishing. Look, if I only move an *inch* a day, that's okay, I'll keep moving. Listen, if all I can do is *point* toward my goal, I'm going to do that, but I'm not quitting.

Resolve to finish. For example, don't get married if you're going to *try* marriage. Marriage isn't a pair of shoes that you try on! People say, "I know if this marriage doesn't work, I can always go and get a divorce." But that isn't the way you go into Marriage.

If you go into marriage with the resolve that you are going to do everything within your power to make it work, then you will have a marriage that lasts. And you'll have more than a lasting marriage. You'll have a marriage that is growing and full of life, because you're working on it. But if we go into marriage saying we're going to *try* it, there's no resolve. The marriage is doomed to fail.

Now I've also heard another dumb thing. A few years ago, the slogan was popular, "Try God." You don't *try* God! God does try us, but we don't *try* God. Jesus said, "No one, having put his hand to the plow, and looking back, is fit for the kingdom of God." (Luke 9:62)

We step into relationship with God with a resolve to remain faithful to Him. "For we have become partakers of Christ if we hold the beginning of our confidence steadfast to the end." (Hebrews 3:14) We need to hold onto our confidence in Christ steadfast to the end! Never quit! Resolve to finish!

Chapter Three
— QUITTING IS A CHARACTER FLAW —

For you have need of endurance, so that after you have done the will of God, you may receive the promise.
Hebrews 10:36

NOW I SAID earlier that the confidence we start with, we've got to take with us. In Hebrews 10:35 we read, "Therefore do not cast away your confidence, which has great reward." Normally we don't start something unless we are *confident* that we can complete it, right?

Nobody can take our confidence from us. *We have to discard it. We* have to throw it away. Nobody else can take away my confidence. Even if I am going through a hellish situation, I can still keep my confidence. And if we do hold onto our confidence, it has great reward!

Hebrews 10:36 goes on to say, "For you have need of endurance, so that after you have done the will of God, you may receive the promise." According to this verse we need endurance or perseverance for two reasons. First, we need endurance to do the will of God. That's what we've been talking about; never quitting! We must endure, persevere and do the will of God.

Second, we need endurance so that *after* we've done the will of God, we may receive the promise. We've endured and done the will of God. But we still need endurance as we wait for the promise. What does it mean that after we've done the will of God we need to endure or persevere that we may receive the promise? It means we never quit!

We may have done the will of God and not yet received the promise. Hang in there. Endure. Don't quit! This confidence comes through the hope that we have through the Scriptures. Knowing and living by the Word of God builds this confidence in us both *to do* the will of God and *endure* until we receive the promise.

Remember, every godly resolve will be challenged. So we must deliberately develop and store up our confidence in the Lord, so that we can endure and go through until we see the finished product.

Why do we need to store up confidence in God? Because, something's going to hit us from the north; something's going to hit us from the south. Something's going to hit us from our family; something's going to hit us from our church. Something's going to hit us from here, from there, from everywhere. But that's okay because, "No weapon formed against you shall prosper." (Isaiah 54:17) We're *confident* in God, now we have to add *endurance.*

Whatever is being hurled at you right now, stay in there my brother; stay in there my sister. Never give up. Never quit! We must decide now that we cannot be defeated and we will not quit! "If God is for us, who can be against us?" (Romans 8:31) That's why our responsibility is to stand fast. *Faith gets us into the supernatural; endurance keeps us there.*

Now, be careful. We can lose our confidence. We can lose our endurance. How do we know when our endurance is breaking down? We listen to our talk. When our endurance is breaking down, we start complaining and getting critical about everybody else.

Endurance is directly tied to our character. God is deeply interested in character, because our character reflects who He is. A trail of unfinished projects is a *character flaw*. A pattern of quitting is a *character flaw*.

First Peter 4:12 says, "Beloved, do not think it strange concerning the fiery trial which is to try you, as though some strange thing happened to you." Peter reminds us that fiery trials are not strange, but common. In fact, trials are supposed to happen. The Bible says, "For to you it has been granted on behalf of Christ, not only to believe in Him, but also to suffer for His sake." (Philippians 1:29)

If we're in Christ, we are going to have a whole lot of trials! God says those fiery trials "try" us. These fiery trials test us to determine the integrity of our character. Will we endure? Or will we give up and quit?

But look at 1 Peter 4:12 and 13:

Beloved, do not think it strange concerning the fiery trial which is to try you, as though some strange thing happened to you; but rejoice to the extent that you partake of Christ's sufferings, that when His glory is revealed, you may also be glad with exceeding joy.

So God is telling us that if we go through this trial (and remember, every season has a beginning and an ending), we'll end up glad and excited.

When we've been through a severe trial and we get to the other side, we no longer remember all the pain and discomfort. We even tell others, "It was worth it all."

Think about this "fiery trial" for a moment. We put metals through fire to bring the impurities to the surface that the natural eye hadn't caught. God will put us through some fiery trials. And the fire of trials will cause our impurities (our character flaws) to come to the surface, making them visible.

Often we may think we already had dealt with that nagging character flaw. We thought it was licked; behind us. And then the trial comes and brings that character flaw to the surface again. That's the purpose of the trial. It exposes our character flaws so that we have to do something with them.

God may bring people into our lives that we think are from hell. God brings people into our lives to bring heat to certain areas of our lives so that the impurities surface that we hadn't seen and no one has been honest with us about. But now, in the heat of this trial, those things come to the surface because they're flaws.

God wants to use us, but our flaw or defect is getting in the way. This is why I say that a lot of us are not promotable. God has been dealing with our flaw, and working on it. But instead of yielding to Him we quit and run.

That's why folks run. They run from church to church. They run from relationship to relationship. They run from job to job. That's why they run. And when they get to their new church, or relationship, or job, because they're new there, their defect is dormant. But we know what's going to happen. God is going to turn the heat up again and try to get rid of that character flaw. God does this because He loves us too much to put us on the market with a defect.

But we forget this and instead of seeing ourselves as God's project, we see ourselves as a *victim*. We think it's all about folks messing with us. "They don't love me. They don't appreciate me. They have it in for me."

I remember one time I was in a church back when I was in the military. I knew a little something and I knew I had a gift, but nobody else seemed to care about my gift. I couldn't get anybody to take notice of me and my gift in that church. And God said to me, "Do you want to know why nobody cares about your gift? It's because you're full of yourself."

Perhaps you fall into the category of those who believe that God has a call on your life. But you complain that others don't recognize you and your calling. Let me ask, "Who called you?"

Remember, those in charge didn't recognize David either. Instead, they thought David's brothers would be the next king. But David's brothers weren't ready and that was proven when the fight came. They were in the hole hiding and David was the only one running toward Goliath.

My point is that God knows how to bring us forward. But He's using trials—often in the form of other people—to burn some stuff out of us. If God called us, God knows how to put us forward.

Chapter Four
— GET PAST THE URGE TO QUIT —

Take this cup away from Me; nevertheless, not what I will, but what You will.
Mark 14:36

THE URGE TO quit will come upon all of us. The urge to quit came upon Jesus. Jesus said, "Father, is there any other way we can accomplish this?" In other words, "I don't want to do this. I don't want to die. I don't want to hang on a cross. I don't want nails going in my hands and feet. Is there any other way we can do this?" Jesus was confronted with the urge to quit.

Nevertheless, Jesus said, "Father, not my will but your will be done." He moved past the urge to quit. And what we have to do is learn how to get past that urge to quit as well.

I look back over my own life, and I've told you already that I was a chronic quitter. But I've learned over the years that if you can hold your course; if you can find the *right* thing to do—this is really important—I mean you can persist in the *wrong* thing. You can be persevering and not quitting, but doing the *wrong* thing. But if you can find the *right* thing—the will of God—and do it and not quit, not cave in, not give up, then there's absolutely nothing that you can't do.

God puts a premium on those who are locked onto his will. When you are locked onto God's will, you're unstoppable! *A believer that understands the will of God and locks onto it is unstoppable.* That person has to *quit* in order to be defeated. There is not a devil in hell that can defeat a Christian who has knowledge and understanding, walks in the will of God, and has the resolve to finish his course. Nothing can stop a person like that—absolutely nothing!

The Bible says in Galatians, "And let us not grow weary while doing good, for in due season we shall reap if we do not lose heart." (Galatians 6:9) "If we don't *lose heart*," can also be translated, "If we don't *faint*." Losing heart or fainting has to do with the mind. Losing heart or fainting is something that takes place in our heads; in our minds.

Hebrews 10:23, also tells us to be steadfast and not to become weary in our mind. "Let us hold fast the confession of *our* hope without wavering, for He who promised *is* faithful." Think about what happens to us when we get weary. When we become weary, or faint, or we waiver, this all happens in our mind. When it comes to quitting, *we quit first in our minds.*

A weak or weary mind will quit in a minute. That is why the Bible exhorts us to renew our minds. We need the Spirit and mind of Christ. Because the devil understands that if he can "ding dong" a person's mind; keep pounding a person's mind with negative thoughts, that person will quit.

By the way, the Bible calls these weapons of the devil "fiery darts." Paul urges us to be strong in the Lord by, "Taking the shield of faith with which you will be able to quench all the fiery darts of the wicked one." (Ephesians 6:16) We take up the shield of faith with our minds. The devil messes with our minds. He knows if he can direct our head (our mind), the rest of the body will follow. Most people quit because they get tired in their minds. So God warns us not to get weary or faint.

The devil makes sure that he sends his mind agents to earth. Those are the voices, thoughts, or fiery darts that constantly oppose what we do. Those thoughts are diametrically opposed to the will of God. These are thoughts that bombard us and tell us things like, "You can't make it. Why don't you give this up? Why are you doing all of this anyway? You know you're going to blow it." Thoughts like these bombard our minds.

I want to show you something in John, chapter 13 to verify this point. In John 13:1, we read, "Now before the Feast of the Passover, when Jesus knew that His hour had come that He should depart from this world to the Father, having loved His own who were in the world, He loved them to the end." Then verse 2 continues, "And supper being ended, the devil having already put it into the heart of Judas Iscariot, Simon's *son,* to betray Him...."

The devil put it in Judas' heart—put these thoughts in him—to betray Jesus. Think about this: Judas was called out by Jesus, he walked with Jesus, and witnessed Jesus' miracles. Judas was anointed of God. It doesn't matter how deep you are in with God. It doesn't matter what your pedigree is or how many degrees you've got. If you don't put a gate and a guard over your mind, you will be vulnerable to the devil because he constantly throws these fiery darts at us.

Somewhere along the line Judas got tired of resisting those thoughts. He got tired of casting out those imaginations and everything that exalts itself against the knowledge of God. Judas got tired of resisting, just like we get tired of it. But we can't get tired of resisting, because when we get tired in our minds, when we let down our guard, and when we give in to these thoughts, actions may follow from which we may have trouble recovering.

Chapter Five

— FIND WHAT GOD WANTS YOU TO DO AND STICK TO IT! —

We do not want you to become lazy, but to imitate those who through faith and patience inherit what has been promised.
Hebrews 6:12 NIV

LET ME OFFER a very simple illustration using a postage stamp. A postage stamp finds its usefulness in its ability *to stick to something until it gets there.* If a postage stamp quits sticking, it becomes useless. A postage stamp that won't stick to a letter is of no use to us.

We are like that postage stamp. All we've got to do is *stick to something.*

I've discovered that the lesson from God for us on this is not deep. All we have to do is just *stick to something* like that stamp. Find out what God wants you to do and *stick* to it.

That doesn't mean we won't have to make some mid-course corrections. The vision doesn't change, but our methodology may change. Just stick to it!

In Hebrews 6:10-12 we read:

God is not unjust; he will not forget your work and the love you have shown him as you have helped his people and continue to help them. [11] *We want each of you to show this same diligence to the very end, so that what you hope for may be fully realized.* ¹² *We do not want you to become lazy, but to imitate those who through faith and patience inherit what has been promised. (NIV)*

Don't just *start* something. Instead, show the same diligence you started with *to the very end*. And we're to imitate, or follow the example of "those who through faith and patience inherit what has been promised." Find somebody who is finishing and do what they do!

If we don't do this, he says we'll become lazy or sluggish. We can be spiritually lazy and we can be lazy about our natural lives—our job, career, family and the like. The lazy person may start something, but they won't finish. The lazy person quits, because it gets too hard.

Let's continue to verses 13-15:

For when God made a promise to Abraham, because He could swear by no one greater, He swore by Himself, ¹⁴ *saying, "Surely blessing I will bless you, and multiplying I will multiply you."* ¹⁵ *And so, after he had patiently endured, he obtained the promise.*

"After Abraham had patiently endured, he obtained the promise." Abraham obtained the promise *after* he had patiently endured. Abraham did not quit. A person who patiently endures does not quit. But we need to *patiently* endure.

We need both patience and perseverance to inherit the promise. Patience and perseverance are like two sides of the same coin. Abraham *patiently endured* and he received the promise.

Patience and endurance are two different things. Endurance and perseverance are synonymous. When we persevere, we don't give up. But it's possible to persevere or endure and not be patient about it. There are a lot of folks who stay with something, but they're not patient about it.

A patient person is one who deals with problems and situations with calmness and self-control. We are not patient when we start saying things like: "I can't believe it! I don't deserve this! Did you see what they did to me?" And then we go off and broadcast it to everybody.

We really don't know how we'll react till the test comes. That is why we need these tests. These tests reveal the flaws in our character. Lack of patience and self-control is a character flaw. If we're going through some trial and we lose our cool and go ballistic, then we know we've got to work on patience. It's not enough to endure if we're not *patiently* enduring.

God cares as much about the *process* as He does with the *product*. We may endure, but if we get there kicking and screaming then we're not representing Christ well. Without this self-control that patience brings, we may do things that we are going to regret. We have a hard time hearing God when we're out of control. When we're impatient and out of control, God may need to keep us "in the wilderness" a while longer until we learn patience. We need to endure with *patience*. "After Abraham patiently endured, he obtained the promise."

Let's look at the endurance or the perseverance side of the coin now. We can be patient, yet fail to persevere. We can be patient and be *passive*. One who is patient, but does not persevere in the pursuit of what was promised won't receive the promise.

What is perseverance? What does a person with perseverance do? A person with perseverance keeps going in the face of severe opposition. They just carry on through, no matter what they are dealing with. Sometimes we may not even recognize that they are dealing with opposition because their MO is to endure it patiently. Their countenance doesn't even change in the face of opposition.

That is a person who understands and has perseverance working for them. They don't care what is happening, they are moving forward. They may be hit with all kinds of issues from every quarter, but they aren't quitting.

A person with perseverance has great stamina. They know how to hang in there. They know how to stick. A person with perseverance refuses to be derailed no matter what obstacle is placed in front of them. They find a way through, over or around that obstacle.

When we are persevering we refuse to give up. When we persevere we are unstoppable.

Our strength doesn't come from what *we* can do. Our strength comes in the consciousness that we have the Greater One living inside of us and there is nothing that anybody can do to thwart us. "If God *is* for us, who *can be* against us?" (Romans 8:31) That promise has to be as real for us as the hand on our arm. That truth has got to be so anchored in us that we persevere no matter what others try to do to us.

"No weapon formed against you shall prosper." (Isaiah 54:17) No weapon formed against us will prosper. "You are of God, little children, and have overcome them, because He who is in you is greater than he who is in the world." (1 John 4:4) God is greater than all others. Our strength does not come from ourselves; our strength comes from God.

"Now thanks *be* to God who always leads us in triumph in Christ." (2 Corinthians 2:14) Oh, we might not look or feel like a winner right now, but God "always leads us in triumph in Christ." In God is our strength. In God we will triumph in Christ.

There are times we're going to say, "I don't know what to do. I just don't know how I'm going to get through this. But my eyes are on You. And I'm expecting You to bring me out of this. I'm trusting You to bring me out of this."

"Lord, in this process, I'm not going to curse You off. I'm not going to hinder Your work by worrying. I'm not going say things to quench Your Spirit. I'm not going to say things that will attract more destruction to me. I'm not going to *bind* this problem to me."

By *binding* a problem to myself, I'm talking about *tying* or *securing* it to me. Every time we talk about our problem in a way that magnifies *it*, we are binding that problem tighter and tighter on us. Don't magnify your problems, magnify your God. Our God is bigger than any problem.

Don't major on temporary stuff that isn't going to be here for that long. Think of those things as disposable. We don't put paper cups in the dish washer do we? Their use is *temporary*. We need to view our problems in the same way. They're *temporary*. They're *disposable*.

Chapter Six
— THERE IS VICTORY FOR THOSE WHO PERSEVERE —

Therefore, my beloved brethren, be steadfast, immovable, always abounding in the work of the Lord, knowing that your labor is not in vain in the Lord.

1 Corinthians 15:58

IN 1 CORINTHIANS 15:58, we read, "Therefore, my beloved brethren, be steadfast, immovable, always abounding in the work of the Lord, knowing that your labor is not in vain in the Lord." God guarantees victory to his children who persevere. When we keep at it and do His will, He guarantees victory. When we have our own agenda and we're not trying to glorify or please God then we're on our own.

Whenever our heart is right toward God and we make decisions in order to please God and glorify God, God will bring into our lives everything we need to cause that to come to pass. He guarantees victory for those who do his will. We "know that our labor is not in vain in the Lord." *Our labor in the Lord is never ever wasted.*

Sometimes, when we try to live for God it seems like everything we do is just useless. But God says, "Be steadfast—don't move off of it."

This Scripture tells me I need to always be *abounding* in the work of the Lord. "Always abounding" doesn't mean cut back! "Always abounding" means to become even more diligent. To abound means to *grow*. And He promises us that our labor is never wasted, it's never in vain.

God says there is a due season coming in which the harvest is going to overtake you. For us to reach our harvest, the only thing between us and our due season is staying steady, persevering, not stopping, not moving, not quitting. We must not quit!

Paul said, "I have learned in whatever state I am, to be content. I can do all things through Christ who strengthens me." (Philippians 4:11 & 13) Paul's faith didn't fluctuate from season to season, because he knew that God is more than enough for any season. Whatever season you are in right now: in the highs, in the lows, or in between, your season will soon change.

We all enjoy being in the high seasons. I like the high seasons when everything is clean, tucked up and turning! In the high season, everything seems to be working out. In the high season, when everything is flowing, you feel like you're Jesus' second cousin!

But when we're in the high season, we know deep down what's going to happen eventually—it isn't always going to be like this.

In the book of Proverbs, God said "Go to the ant, you sluggard! Consider her ways and be wise." (Proverbs 6:6) God tells us to go to ant school to learn something from them. What we can learn from the ants is that they store up food for winter during the summer. In the high season, the ant is preparing for his next season. When the ant is harvesting, he is not just sitting back consuming everything he's set aside.

God said "Go to the ant, you sluggard!" A sluggard is a lazy person. We become sluggish and lazy when we back off our spiritual activity and we're no longer diligent. God says, "Go to the ant, you sluggard! Consider her ways and be wise, [7] which, having no captain, overseer or ruler, [8] provides her supplies in the summer, *and* gathers her food in the harvest." (Proverbs 6:6-8)

No one has to instruct the ant to do this. It's inborn. The ant knows it's not always going to be like this. The ant knows that in the winter when there isn't any cornbread to be found, it won't be a problem, because he's got plenty stashed away.

Some of us think that our finances are always going to be flowing like they're flowing right now. We are not thinking about the fact that this season may not always be like this. Oh, we may not lose our job or anything that serious, but something will come down the pike that will cause some unexpected interruptions in our cash flow.

We must not spend everything we get at harvest time. Some of what we harvest is bread for food, but some is seed for the next season. Go to the ant. Go to ant school. Prepare now for lean times coming. We're not trusting in the money, we're trusting in God. But God told us to go to ant school and prepare.

We've talked about the high times, but there are also low times. They will be temporary too. Seasons change, but we've got to learn how to maximize every season. I like the low seasons because God always builds something in me. I don't *enjoy* the low seasons, but I *like* them because they force me to dig deep spiritually. And that digging deep spiritually is not something for that season only. After digging deep spiritually in my low seasons, now I have something to take with me through the high seasons as well.

We must not forget from where God has brought us. We must not forget where we came from. We need to remember what God has saved us from. We need to remember how we've grown in Him. We all have memories that are like pictures of how God has changed us; memories that cause us to bow in humble adoration before Him.

Here is one of those pictures or memories for me: I'll never forget. I was sitting in the Victory Terrace Projects, 671 Magnolia Avenue. I was about 16. I was lying on my bed in the west bedroom listening to the Delfonics. My brother and I had stolen a record player. It was one of those record players with a big speaker that you put on top. We didn't have any money, and we took one so we could listen to music. But I vividly remember as I lay there. I vowed that I would never ever in my life, live like this again.

Here I am 36 years removed from that time. And when I go to my garage and pick which car I'm going to drive, I say, "My Father, I thank you." And when I go to one of my refrigerators to get something to eat, I say, "Thank you, Father, that I don't have to rush home before my brother or family to make sure there's something left for me to eat." That incident still goes through my mind after 36 years. I can't forget it. Memories like that keep us going. Remember and rehearse them.

Chapter Seven
— WE HAVE THE WORLD OVERCOMER RESIDING INSIDE OF US —

For everyone born of God overcomes the world. This is the victory that has overcome the world, even our faith.
1 John 5:4 NIV

"THEREFORE, MY BELOVED brethren, be steadfast, immovable, always abounding in the work of the Lord, knowing that your labor is not in vain in the Lord." (1 Corinthians 15:58) This Scripture will encourage you over and over again. When you are going through a low season and you can't even see how you'll make it, if you abound in the work of the Lord, God guarantees He will come through for you. Your labor is not in vain.

If you are a married couple and one of you has your mind set on doing the will of God, whether your spouse wants to do the will of God or not, you have a choice. "Be steadfast, immovable, always abounding in the work of the Lord." This won't be easy. But no matter how difficult it gets—even insanely difficult to do the will of God—you have to realize that God is completely dependable. When the pressure is on, then we find out whether we really believe that our "labor is not in vain." When the heat gets turned up, we learn that God is completely dependable.

What are we going to do now that everything is going wrong? We're going to do what we've always done. That should be our answer. We're going to do what we've always done. We're going to pray to God. We're going to depend on God. We're going to "be steadfast, immovable, always abounding in the work of the Lord, knowing that our labor is not in vain in the Lord."

Let's look at another Scripture that tells us when we do the will of God we can't be stopped. In Acts 5:38-39 we read, "And now I say to you, keep away from these men and let them alone; for if this plan or this work is of men, it will come to nothing; but if it is of God, you cannot overthrow it—lest you even be found to fight against God."

This incident took place after the apostles had been thrown in jail for preaching the Gospel. Then that night, an angel of the Lord released them and told them, "Go, stand in the temple and speak to the people all the words of this life." (Acts 5:20) The chief priests and the Sanhedrin arrested them again and wanted to kill them.

Gamaliel reasoned that if the apostle's message was of God, then it didn't matter what the Sanhedrin tried to do, they wouldn't be able to stop it. Gamaliel was talking about the apostles and their ministry, but it is true with us as well. If what we're doing is the will of God, there is no man, no system, no group, there's nothing that can kill what God has birthed. Nothing can stop us!

Perhaps you're in a situation and you're dealing with some difficult things and you say, "I'm going to handle this God's way." You are the only person that you can control in your situation. You can't control anybody else. The only person you have control over is *you*.

Now if our goal is something we dreamed up that we're promoting because we want to be somebody special and make a name for ourselves, that dream can be short lived. But according to this passage, if we are doing the will of God, there is nothing that can stop us, *if we persevere.*

In spite of how other folks are acting, God said, "I will watch over you, I will keep you." We can't control what other people do, that is something they've got to do for themselves. Don't let folks get under your skin. We cannot allow what others say and do to disturb our peace and deter us from doing the will of God.

Allow me to share a personal example here. Jesus said, "I will build My church, and the gates of Hades shall not prevail against it." (Matthew 16:18) "I will build *My* church." At one point in my ministry, I realized *I* was trying to build a church for *Him.* Then I heard Louis Greenup, a preacher from New Orleans, Louisiana. He was a pastor and he said the same thing that I had—that *he* was trying to build a church for Jesus.

Louis said, "This is the truth, I was committing adultery." What he was getting at is this. Louis said:

> *I was doing with the church what I'm supposed to be doing with my wife. God told me to feed the church, be an example to them and when I get done with that go home and be with my wife. But here I was trying to counsel everybody, always running here and there trying to make sure everybody was okay and I was trying so hard to build His church. But God said to me, "You're supposed to be making sure your wife is okay. You're supposed to be nourishing your wife. You're*

supposed to be fellowshipping with your wife." God reminded me my wife is my bride.

Then He said to me, "Leave My wife alone. I'll take care of her. Feed my flock and be an example to them, love them, and pray for them. But you take care of your wife."

After hearing Louis, I thought, "I need to go home and be with my wife." Jesus is not into wife swapping. Sometimes we pray, "Lord, bless my wife!" But God says to us husbands, "No, *you* bless them."

A lot of pastors and others who are involved in ministry are committing wife swapping. They are at church all the time. They get all burned out and have no home life because they are swapping wives. Jesus says to us, "You leave my woman alone! You go home and be with your woman." I was trying to build the church for Jesus. We've got to let God do what He does while being faithful doing what He has called us to do.

We cannot fail when we are doing the will of God. But many people don't know how to apply that to their work. What does God say about my job in terms of doing His will? If we don't know the will of God regarding our job, we're going to treat our job wrongly. If we have a wrong approach to our job, God can't do what He wants to do through us in that job.

We need to understand what God says about working and laboring. How does God tell me to treat my employees, or my boss or supervisor? We've got to approach our jobs according to the will of God.

We need to have a proper perspective about our jobs also. I may lose my job. The company might shut down, but I won't lack anything. As soon as one door closes, another one opens. This isn't a big deal for us. Why? – Because we did this thing God's way. We go to our place of work to be a blessing. We don't go to work just to *take*. We're not just *consumers*. We go to work to add something; to contribute.

Whatever we do according to the will of God, we can't be stopped. But we must persevere. In John 16:33 Jesus said, "These things I have spoken to you, that in Me you may have peace. In the world you will have tribulation; but be of good cheer, I have overcome the world."

Why did Jesus speak these things? – That in Him we may have peace. In this world we *will have*, not *might have*, "in this world you *will have* tribulation," trials, times of great frustrations, tests, misunderstandings, "but be of good cheer." Why?— "Because I have overcome the world." The Amplified version says, "For I have overcome the world. [I have deprived it of power to harm you and have conquered it for you.]"

What Jesus is saying is this: we're going to go through all these tribulations— and He said, "I have overcome the world." The Apostle John also wrote, "You are of God, little children, and have overcome them, because He who is in you is greater than he who is in the world." (1 John 4:4) Consider these truths from this perspective. Jesus Christ lives in us and He is greater than all. If Jesus has overcome the world, and He lives in us, then we too are overcomers. That is why He said, "Be of good cheer."

John continues this theme in 1 John 5:4, "For everyone born of God overcomes the world. This is the victory that has overcome the world, even our faith." (NIV) Whatever the world throws at us; whatever turmoil is in the world; whatever we encounter in the world; this is the victory that overcomes the world, even our faith.

We have a consciousness that we have the World Overcomer residing inside of us. Jesus says, "Be of good cheer, be conscious of the fact that I am in you and I will take you through whatever comes at you, for I have overcome." In Galatians 2:20 Paul says, "I have been crucified with Christ; it is no longer I who live, but Christ lives in me; and the *life* which I now live in the flesh I live by faith in the Son of God, who loved me and gave Himself for me."

The truth that Jesus, the Overcomer, resides in us has to be real for us. He lives in us and He will bring us through whatever we're experiencing. Because He is the Overcomer, we are overcomers.

Why do we get filled with the Holy Ghost, to speak in tongues? No! Jesus said I'm going to send you power from on high. Why do we need power? – Because without His power, without His Holy Spirit, we'll get sucked into the world like everybody else.

The Holy Spirit gives us power to overcome the world and while we're overcoming, we can help some other folks. We don't have to quit! We don't have to take the easy way out.

Think of persevering in terms of running. Some folks never get their second wind. Runners talk about getting their second wind. But some folks never get their second wind, because they never exhausted their first one. We don't get a second wind till we use up the first one. But we can't give up.

Ordinary talent, with extraordinary perseverance can accomplish incredible things! But we can't stop. Pick it up again. Pick up that fight again. Go for it! Don't back down. Don't quit!

Chapter Eight
— CULTIVATE A FINISHER MINDSET —

Be transformed by the renewing of your mind, that you may prove what is that good and acceptable and perfect will of God.

Romans 12:2

WE LIVE IN a society where commitment, a real commitment, doesn't mean a whole lot. When we become born again, our character doesn't change overnight. For instance, one of the things I said earlier is that if we have a string of unfinished projects in our life, those things will testify against us according to Luke 14. We started, but weren't able to finish.

If we assess our lives and find there's a whole lot of things that got stamped "unfinished," "unfinished," "unfinished," then we've got some character deficiencies. Something's wrong and that deficiency needs to be addressed. Otherwise, that character deficiency will keep us far out of the will of God.

In a previous chapter, I stated that if we quit a lot, it desensitizes us to commitment and we develop a quitter's mentality. And with a quitter's mentality, we'll quit anything. We'll even quit God. A lot of times we don't even think about what we're doing when we quit something.

We've developed a pattern. We've established a script, "If I don't finish, it's okay. If this thing doesn't work out, it's okay."

A lot of people go through life with that mentality. This attitude may slop over to our job as well. My wife and I have a relative. This relative is a very intelligent, beautiful, awesome woman. But in *every* job she had, she had a *crazy* boss. According to her, in every job she ever had, her boss had it in for her—in every job!

Finally, I had to say something to her. I said, "You know, the only common denominator in all those scenarios is *you*. Maybe it's time to consider that maybe something's wrong with you. Could it be that there's a character flaw here?" Some of us may be in the same situation as my relative. If so, it's time to go to work on us.

I was a quitter. I went from being a quitter to a finisher. Everything changed once I started finishing. The experts tell us that in goal setting we need to begin by setting a goal we can reach. Don't set a goal so far out there that there's no way you'll ever finish. Go after something that's attainable. Why? – Because something happens when we experience accomplishment.

We are wired to move forward and to progress. Every human being was made to progress, to increase. "Be fruitful and multiply." The desire to accomplish something and to grow is inborn. When we complete something and can stand back and say, "I did that," something happens to us. We gain confidence. We almost want to say, "What other universe can I conquer?" This is why we've got to get in the habit of finishing.

At this point, I'd like to describe how we do this. I'm going to provide some practical things to help us develop some internal stamina. Most of us know what to do. We just don't do it long enough.

What causes people to quit? Why do we find quitting so easy? As I said, when we become born again, our character doesn't change overnight. This is important to realize, because sometimes we think that when we are born again, everything changes. There's a song out there that says, "I looked at my hands. My hands looked new. I looked at my feet and they did too." I understand what the songwriter means, but in reality this is not true. Coming to Christ did not give me new hands. If I wore dentures before I came to Christ, I didn't get real teeth once I became saved.

We can say the same thing about our character. If we were quitters before we got saved, we're likely to remain quitters after we're saved. We've got to deal with quitting, because it's a character issue. God said in Isaiah 55:9, "For *as* the heavens are higher than the earth, so are My ways higher than your ways, and My thoughts than your thoughts." In other words, when we come into the kingdom of God, we have inferior thinking. If we don't deal with that thinking, we'll still be saved all right, but we may be walking in defeat because we've got a quitter's mindset.

In Romans 12:2, Paul said, "Be transformed by the renewing of your mind, that you may prove what *is* that good and acceptable and perfect will of God." If we don't renew our minds, we could be saved and on our way to heaven, but we'll still quit at the drop of a hat.

Why do people quit so easily? In 2 Timothy 3:10 Paul writes:

But you have carefully followed my doctrine, manner of life, purpose, faith, longsuffering, love, perseverance,[11] *persecutions, afflictions, which happened to me at Antioch, at Iconium, at Lystra—what persecutions I endured. And out of them all the Lord delivered me.*

In Paul's doctrine, manner of life, purpose, faith, longsuffering, love, perseverance, persecutions and afflictions he *endured. And the Lord delivered him out of them all!* Wow! Let's reflect on this passage and what it means for us personally.

First of all, we understand right up front, we're going to have some difficult times. In verse 12 of that chapter Paul says, "Yes, and all who desire to live godly in Christ Jesus will suffer persecution." Why do people quit so easily? One reason is persecution, affliction, and trials. The Bible teaches us that we cannot escape persecution if we're living godly in Christ Jesus.

This fact only leaves one thing left for me to do: and that's to prepare for persecution and trials to come. I'm going to have to deal with it. That's why Paul said, "I glory in tribulation." That's why James said, "Count it all joy when you fall into various trials." Paul said, "You don't pray it away. You don't get somebody to lay hands on you. You don't go to counseling." He said what? "You endure it."

Where does persecution come from? It comes from several places. Persecution sometimes comes from *sinners*. It sometimes comes from *saints*. It sometimes comes from *self*. And it sometimes comes from *society*.

And how do we respond? We endure. We outlast it. That's what Paul was saying. We outlast persecution. Nobody enjoys persecution and our first response is to find release from it. But we are to endure persecution and not quit.

When I was growing up, dogs ran around free in the community. Some of these dogs liked to chase cars. They'd wait alongside of the road until a car came, then they'd run out after the car and chase it for a while. When they realized they couldn't keep up with the car, they'd come back and wait for the next car.

I often thought to myself, "What's that dog going to do if it catches a car?" And what I realized is that those dogs are just like people who persecute. I think the dog's greatest satisfaction would be to get the car to *stop*. And I believe that the people who persecute us would like nothing better than to get us to stop.

But would the persecution end there? No! Let me tell you what would happen once we stop. They'll start criticizing us that we started something we couldn't finish.

We have to realize that persecution is designed to get us to quit. That's what it's all about. For instance, when my wife and I got saved in 1983, we got saved for real. But our friends and acquaintances talked a lot about us. They'd say to us, "You all think you're all this. You all think you're all that. Why do you to go to church so much? I can't believe you all. You are going crazy over that Bible. Why do you all have to be together all the time?"

Some of these folks who were saying these things were church goers. I'm telling you, persecution comes from the sinners, the saints, society and self. They talked about us incessantly. I can't believe that they tried to make me stop. Who do they think they were? I realize now, thirty years later, I'm so glad I didn't quit.

I want to look at something that reveals the purpose of persecution as God wants us to see it. We find this in Mark 10:29-30:

> So Jesus answered and said, "Assuredly, I say to you, there is no one who has left house or brothers or sisters or father or mother or wife or children or lands, for My sake and the gospel's, [30] who shall not receive a hundredfold now in this time—houses and brothers and sisters and mothers and children and lands, with persecutions—and in the age to come, eternal life.

When we receive back "a hundredfold," this means we receive maximum, or optimum return. Don't cheapen this text by trying to apply it to money. Jesus wasn't talking about money. He's talking about relationships. Those who have given up, "houses or brothers or sisters or fathers or mothers or children or lands for My sake," will receive back now a hundredfold of the same. That's God's best. That's maximum return. That's God bringing something wonderful into our lives. On top of this, He promises us, "And in the age to come, eternal life."

But meanwhile, we're going to have to deal with harassment and persecution.

I mentioned earlier that persecution is designed to get us to quit. Mark 4:16-17 gives us an example of this:

> These likewise are the ones sown on stony ground who, when they hear the word, immediately receive it with gladness; [17] and they have no root in themselves, and so endure only for a time. Afterward, when tribulation or persecution arises for the word's sake, immediately they stumble.

Without deep roots, when tribulation or persecution comes, we'll stumble or become offended and quit.

Now, it may surprise us *where* persecution comes from, but once we get over that, we can say, "Well okay, I knew it was coming. I just didn't know it was coming through him." Yeah, persecutions and tribulations will come.

No man or woman has ever reached a higher level of success without persecution, hardship, and tribulation. No one ever has. But it's possible to take hardship, hard times, and tribulation and let them propel us past the urge to quit.

Chapter Nine
— THE UNSTOPPABLE SPIRIT —

*I have fought the good fight, I have finished the race,
I have kept the faith.*
2 Timothy 4:7

HARDSHIPS, TRIALS, PERSECUTIONS and difficulties will come. They're inevitable and part of the Christian life. We cannot spend our life tip-toeing around difficult times, or running from them. We cannot build a wall of pillows around us where everything is comfortable. If all we want to do is stay where it's comfortable, we'll never stretch, we'll never experience anything.

I'd like to introduce you to what I call the *unstoppable spirit*. This unstoppable spirit is in every believer. Overcoming is in our DNA. It's in there. There's an unstoppable spirit residing within us. Paul tapped into this unstoppable spirit and could say, "I have fought the good fight, I have finished the race, I have kept the faith." (2 Timothy 4:7)

Few people reach past pain, pressure, discouragement, and the urge to quit so they can tap into that unstoppable spirit, but you can. I said earlier that we won't get our second wind if we don't exhaust our first wind. If we can't reach past pain, if we can't reach past disappointment, if we can't reach past hardness, if we can't reach past all that, we'll never tap into that unstoppable spirit.

The Bible says things like, "God always leads us in triumph in Christ." (2 Corinthians 2:14) That's part of the unstoppable spirit. It's in our DNA. 1 John 5:4 says, " For everyone who has been born of God overcomes the world." (ESV) Whoever is born of God ..." does what? "Overcomes the world."

So where is persecution? Where is the urge to quit? Where is temptation? Where is all that? That's a part of the world. If we don't face these things head-on; if we whimper every time we're confronted; if we turn and try to find comfort in something else every time something doesn't go our way, then we turn into quitters—losers! Just because somebody doesn't come through for us, that doesn't rock our world, why? Because we're overcomers! We have to tap into that unstoppable spirit that God provides.

Think about it. There's only very little you and I can control. But one of those things we can control is our emotions. We can't control what other folks do. No matter who we are or how hard we try, the bottom line is we can't control what other people do. That's why God said, "Don't be putting trust in man. You don't know what they're going to do." (Psalm 146:3 paraphrase)

When the pressure is on, then we have to say, "Now I know what I'm really made of because if I try to find an excuse; if I try to find comfort in somebody else, or try to get somebody else to do this for me, I will never ever grow to a place where I can be a conqueror and tap into that unstoppable spirit."

If we're blaming somebody else all the time—and we do play that blame game—we'll never win. We blame other people, or the job, or the house. "If it wasn't for my kids; if it wasn't for my in-laws; etc., etc." Come on, we've got to grow up!

That unstoppable spirit is in there. As a believer, it's in us, but it requires pressure to release. We don't get olive oil without squeezing the olive. I can't go on to the next level until I master this one.

God asks us, "Will you be there? Will you persevere?

A lot of us get right up to the edge. Sometimes we're just a step or two away from breakthrough, and then we quit. But Jesus, who lives inside of us, has overcome the world. Christ, the hope of glory, lives within us! We need to renew our minds and have a Bible consciousness of the Spirit of God living in us, or we'll never tap into that unstoppable spirit.

Look again at Hebrews 10:35-36, "Therefore do not cast away your confidence, which has great reward. [36]For you have need of endurance, so that after you have done the will of God, you may receive the promise."

Christ is ever changing our character. He is changing us from being a quitter to a finisher. But we have need of *endurance*, so that after we have done the will of God, we will receive the promise.

To endure or persevere is something we have to *do*. We don't receive endurance in the prayer closet and change overnight from a quitter to a finisher. Finishing is not an accident. Finishing is not a prayer we pray. We don't just pray and become successful. *Change, permanent change, is a product of human effort, sustained by divine power.*

I'm not trying to minimize the importance of prayer, but some people get the wrong idea about prayer. They think, "All I have to do is pray this prayer and I'll be transformed from a quitter to a finisher." No, it doesn't happen like that! There's some work involved. Finishing takes effort.

Most folks don't want to work. They want somebody else to do the work for them. It takes work to change. It takes effort to go from that person who always has their hand out begging, to become someone who can give a hand out.

You can do anything you want to. If you're willing to hold your course and not quit, you can begin to predict your future. I'm telling you the truth. "After you've done the will of God, you have need of endurance." Persevering is something you've got to do. You can't just talk about it.

The Bible talks about the law of life in Christ Jesus that frees us from the law of sin and death. There are laws: divine laws, laws of success, and laws of just about everything. If you learn how to lock into the law, there's nothing you can't accomplish!

Years ago, I became acquainted with a man named Les Brown. Les and his twin brother were adopted when they were six weeks old by a single woman with little education or money. Early on, Les was mislabeled as a slow learner, which severely damaged his self-esteem for years to come. He had no formal education beyond high school.

I knew Les Brown when he was sweeping a radio station. I knew Les Brown when he was standing on the street corner and nobody was listening to him. Today, Les Brown is a motivational speaker and entrepreneur earning as much as $40,000 for a single speaking engagement.

Les wrote a book entitled, "Fight for Your Dreams: the Power of Never Giving Up!" Fortunately, Les didn't listen to what others said about him. He didn't let his circumstances dictate his future.

What other folks think and say have no bearing on us unless we let it. Les had that unstoppable thing in him. He was the first one I heard use that term "unstoppable."

What are people saying about us? What is it that keeps us from moving forward? In what area of our lives do we keep quitting? When are we going to stop falling for the lies that we can't finish? After we've done the will of God, then we need endurance, then we need patience.

Think about something in your life that you have accomplished. Now think about how you stayed with that task until you finished it. No doubt you experienced times when you wondered, "I don't even know if this is worth it." But you stayed with it.

Maybe a situation occurred and you said, "Man this is so hard, I don't know what I am going to do." But you stayed with it, you overcame. You conquered. There was human effort at work there. But before I'm accused of pure self-development and attaining one's full human potential, let me explain.

In the church, we have a tendency to spiritualize everything and simply to commit everything to God in prayer. Now, I'm a big believer in prayer. But many times we pray asking God to do what we're supposed to do. We treat God like an errand boy, asking Him to do things for us that He wants us to do.

Let's use a jet aircraft as an example here. There are many different kinds and sizes of aircraft. Each plane comes with a specific set of operating instructions. If we're going to fly such an aircraft, we need to know a variety of things about operating that specific airplane. One of the factors that comes into play is the length of the runway. In order to take off or land, an airplane requires a certain length airstrip.

Let's say we have a Boeing 737-600 aircraft that requires 8,016 feet of runway at its maximum takeoff weight. In order for this metal bird to overcome the law of gravity, it needs to accelerate over the 8,016 feet of runway. Only then will the law of lift supersede the law of gravity.

Suppose we're sitting at the start of a runway that is plenty long enough for that 737-600 jet to take off. We gain clearance from the tower and we push the throttle forward. We're barreling down the runway gaining more and more speed, but at 6,000 feet of runway, we get cold feet so we back off on the throttle. Why didn't we take off? Because we quit before we got to the place where the law of lift could kick in.

Every time we quit, we lose something. After we quit, we find out that (like that 737 jet) if we had held on for a little while longer we could've gotten airborne.

There's another important analogy here about airplanes: they work best when we take off *into* or *against* the wind. There's more resistance, but the headwind produces more lift. Flying into opposition takes the pilot higher faster.

This is why Paul said, "We also glory in tribulations...." (Romans 5:3) Why? Because the more trouble I have to deal with, the higher and faster I can get up over that thing. That's why the Bible says, "And we know that all things work together for good to those who love God, to those who are the called according to *His* purpose." (Romans 8:28) "*All things* work together for good to those who love God."

Therefore, trials and oppositions are not a problem for me. What are they? Something that's going to take me higher, faster. God says, "Listen, tribulation is going to happen, but it's going to work for your good."

In the face of tribulation, our testimony is not going to change. Our joy is not going to change. Tribulations are just a part of living right here. We should expect some opposition. We don't like it, but we know it's working for our good. And because it's working for our good, we're going to endure it, and we expect to come out on top.

"Many *are* the afflictions of the righteous, but the Lord delivers him out of them all." (Psalm 34:19) During a crisis, we learn a lot about ourselves, about God, about the devil, and about our friends. In crisis, a whole lot of things emerge. But one thing we can be sure of, a crisis will work for my good if I work the law of lift and keep going down that runway. The Lord said, "Don't get weary in well doing."

Chapter Ten
— SET YOUR WILL TO WIN! —

No one, having put his hand to the plow, and looking back, is fit for the kingdom of God.
Luke 9:62

WE NEED TO get to the point where we detest quitting! We don't talk about quitting. We detest quitting so much that we don't even want to hang around quitters. We don't want their quitting to rub off on us.

But we've got to deal with something here, because some of us have stopped something that God ordained and we think that's okay. That's not okay. "For the gifts and the calling of God *are* irrevocable." (Romans 11:29) God meant what He said when He told you to start. Pick that thing back up. You were so zealous for God. You declared, "God called me to do this." Now finish it!

A lot of people think that in order to change they need some kind of vision from heaven. But all we need is the right motivation. Over 27 years ago, my wife gave me that motivation to change in the form of an ultimatum! She still has my letter today to prove it. She gave me an ultimatum and guess what, I changed just like that.

In that letter, I wrote down how I was going to change and I told her, "Baby, if you stick with me, you'll be so glad you did. If you stick with me, you'll be the happiest woman on earth. You won't lack for anything." I recorded that in a letter. I wasn't even saved, but I told her I'd change and become everything she ever wanted in a man.

Now I'm not bragging, but I want you to listen to somebody who knows what he's talking about. We make all kinds of excuses for not changing and not finishing. But you can change if you want to. I went from being a quitter to a finisher, one who can hold his course. Something happened and my whole life was transformed.

That's why I'm so excited to tell folks that they can do this. *You can do this!* Anybody can! Back then, I wasn't even saved. I didn't have the Holy Spirit yet, but I resolved to change.

I gave you what I believe is the number one reason that people quit. The number one reason people quit is because of persecution or difficult times.

I don't want to leave the impression that the devil is behind all of our trials. That's not so. Where do trials, persecution and difficult times come from? I think there are basically four reasons that crisis comes into our lives:

- Satanic attack
- Human error (my messing up)
- The call of God on our lives makes us a target
- The times we're living in

Those seem to be the four major reasons why we have to go through some of the hell that we go through on earth.

Consider first satanic attack. In the book of Job, we find the account of Satan attacking Job. He attacked his family, his possessions, his finances, and then finally his physical body.

Job didn't have the advantage that we have of knowing the source of his troubles. Even so Job said, "Though He slay me, yet will I trust Him." (Job 13:15)

We've already looked at 1 John 4:4, "You are of God, little children, and have overcome them, because He who is in you is greater than he who is in the world." Yes, we have an enemy. But greater is Christ in us than he who is in the world—the devil.

The second source of trials is human error. We can do things that create crisis, storms, persecution, and affliction. One of the primary ways we do this is by disobeying God. Jonah once found himself in the belly of a great fish, why? The devil didn't do that. No, it was because Jonah blatantly disobeyed God.

This is why in Deuteronomy, Moses told Israel, "If you obey God, all these blessings will come upon you. But if you don't obey, all these curses will come upon you and overtake you." (See Deuteronomy 28:1 and 15.) Human error and disobedience to God can cause crisis. We can make decisions that will bring about crisis and persecution. That's not the devil. It's us.

A third source of trials is the call of God on our lives. In Acts 20:22-24, Paul said:

And see, now I go bound in the spirit to Jerusalem, not knowing the things that will happen to me there, [23] *except that the Holy Spirit testifies in every city, saying that chains and tribulations await me.* [24] *But none of these things move me; nor do I count my life dear to myself, so that I may finish my race with joy, and the ministry which I received from the Lord Jesus, to testify to the gospel of the grace of God.*

Paul said, "Chains and tribulations await me. But none of these things move me." Why? He knew these things were all designed to stop him and get him to quit. But in spite of all those tribulations, Paul purposed to "finish his race with joy."

When God puts an assignment in your life like that and sends you out to impact others for Christ, we know that all hell will break loose. But that's okay. Paul was compelled by the Spirit to go and he was obedient.

When my wife and I first moved to Anchorage to start the church that I pastor there, we knew that God had sent us here. We were on assignment from Him. But within our first few weeks all hell broke loose! First, a lady hit us with her car—almost killing us—and she lied to the police. We had to go to court over that and eventually won.

Then shortly after the accident, my wife had a miscarriage. Then we received a letter from the government telling us to find a lawyer, because they were going to take everything away from us. All this happened within the first 60 or 90 days after our arrival. These attacks came in direct response to our obedience to follow God to Anchorage. This was an attack on our call from God. These attacks were aimed at getting us to quit and get out of here.

The final source of our trials and tribulations has to do with the times we're living in. For instance, in past years the recession hit us hard. The price of gasoline and nearly everything else except homes went up. People lost their jobs. Bonuses and raises were cut short or eliminated. Money has been tight. None of those things resulted from anything you or I did. These things are simply signs of the times.

There are things that are beyond our control that can happen in society, or in our lives, but that's not a reason to quit. God says, "My commitment to you doesn't waver." When the brook dried up for Elijah, God kicked in another plan. He sent Elijah to a poor widow to sustain him with food and drink.

Be assured that there is a reason for every storm and challenge in our lives. We don't always know that reason up front, but one thing is for sure, God said, "Many *are* the afflictions of the righteous, but the Lord delivers him out of them all." (Psalm 34:19) It doesn't matter *why* the tribulation happened. It's the will of God for me to go through it, and to come out victorious—to learn from it, to get higher, stronger, more confident, more anointed to do what God called me to do.

We have talked about the devastation of not finishing. If I want to be a finisher, I have to set my will to win. We need to set our will. Our will controls what we believe. Our will controls whether we will hold to a course. Our will determines our being and the consistency of our character. The strongest, most dominant authority in our lives is our will.

Now we don't like to talk a lot about our will, because we want God to do everything. But God won't do what we're supposed to do. And there's also a battle for our will. The devil wants our will and God wants it, because whoever has it, can find expression through it. There's a battle for your will and mine.

God is hard after our will. The devil *can't* violate our will and take it and God *won't*. I've got to give my will to Him.

If we're going to learn how to be finishers and how to press through and how to see it through to the other side, we're going to have to be able to set our will.

Determination is a product of the will. A lot of folks make decisions but they don't back those decisions up with determination. When we start something, we're going to have all kinds of opportunities to quit, to find the easy way out, to water down our commitment, to back off, and to put it on hold. But determination says, "I don't care what they say, or do. I must press forward and persist until I succeed."

Here's what it means to set our will. "It means to have an uncompromising resolve, to stay on course until we reach a predetermined goal." What is your goal? Whatever it is, if you want God's help, you're going to have to set your will.

Will power is the strength of our will. Think about a time when you were presented with a challenge or some inspiring information that prompted you to say, "You know what, I want to do that. That's what I've been looking for!"

You may have attended a seminar or a program and the information you received stirred you. Perhaps this book is taking you to a place where you say, "You know what, I'm going to stop quitting!" The strength of your will is tied to that information you received. Whatever information initially caused you to set your will, you'll want to continually remind yourself of that information. Rehearse it, allow it to continually flow into your life to strengthen your will until you reach your predetermined goal.

Organizations like Weight Watchers, Alcoholics Anonymous, and others like them have what they call "support groups." There's a constant flow of information and encouragement to stay with it from those in these support groups. Those who attend and keep immersing themselves in this information and accountability tend to stick with it and succeed.

As another example, every year in January an interesting phenomenon occurs. All the athletic clubs and gyms fill up. You can't even find a parking spot! But come around there March, April, and May and there's plenty of room at the athletic club! It's continual information and the support and encouragement of others that keeps the will strong.

This is how to finish. God made us and knows what makes us tick. That's why He said, "Meditate on the Word," and "Keep meeting together." But if we cut off the information and remove ourselves from those who would encourage us, we'll become weak and won't be able to stand.

What I'm sharing is not rocket science. It's not a mystery. If all we do is just come to church, but don't do anything when we're away from here, we won't hold our course for very long. It's just a matter of time before we quit.

Chapter Eleven
— THE MOTIVATION OF PAIN & PLEASURE —

Do not be deceived, God is not mocked; for whatever a man sows, that he will also reap. For he who sows to his flesh will of the flesh reap corruption, but he who sows to the Spirit will of the Spirit reap everlasting life.
Galatians 6:7-8

SOME YEARS AGO I worked with a woman named Debbie who had just been diagnosed with cancer. She called the church asking to see me. I requested that when she came, she would come ready to tell me what she believed—what she was trusting God for in her situation. We met and talked and I saw that she was resolute to trust God through her cancer.

So all I had to do was help her stay on course, because she was suffering in pain—terrible pain. Pain is never easy and it's our natural tendency to want it over now. Later, when Debbie was going through chemotherapy, I remember we talked again. I asked her, "Debbie, are you all right?" And she said with conviction, "Yes, I'm all right."

I assured her, "Yeah, you're going to be all right. It's going to be good." And Debbie held her course. She never changed her language or positive outlook and by doing so, she changed her circumstances. She didn't allow her situation to determine her focus or attitude. Debbie persevered.

But most people don't walk by faith like Debbie did. Most people walk by sight. We just get caught up in what we see with our physical eyes. Faith means you see what's happening with your physical eyes, but you know there's another realm.

How do we know whether we're walking by faith? We listen to our words. What are we saying? Are we saying things like, "I don't see how I'm going to get past this!" That isn't faith. "I don't know when this is ever going to change!" That isn't faith. "I feel helpless. I don't know what we're going to do next!" That isn't faith. "But I can't wait till God comes through!" That isn't faith.

In 1 Corinthians 7:37 Paul says, "Nevertheless he who stands steadfast in his heart, having no necessity, but has power over his own will, and has so determined in his heart that he will keep his virgin, does well." Such a man "has power over his own will, and has so determined."

Now this passage is in context of sexual abstinence, but the principle is universal that we've got power over our own will. And Paul says here that this man who has power over his will, "does well." The sexual drive is one of the most powerful drives a man or woman has. So if a man or woman has power over their own will in the area of sexual abstinence, then surely we can maintain power over our will in other areas of life!

Again, the text says, "This man has determined and he has power over his own will." You can determine in your heart and set your will to change today because you have the power to do that. If you have some destructive behavior, or if there's something that's keeping you from obtaining what you know God wants for you, you can set your will today.

In Romans 7:22-23, we find a principle at work. Paul says there, "For I delight in the law of God according to the inward man. But I see another law in my members, warring against the law of my mind, and bringing me into captivity to the law of sin which is in my members." Notice that Paul refers here to three laws: the law of God, the law of his mind, and the law of sin.

A *law* is a system, or an orderly way that something works. Something that is a *law* will work every time. So Paul says, "There's a law of God, a law in my members or the law of sin, and the law of my mind." What I want to focus on is this *law of my mind*, because we're talking about setting our wills. If I'm going to set my will, I must understand how this law of my mind works and then systematically apply it.

I wrote in a previous chapter about the law of lift overcoming the law of gravity in an airplane. Similarly, Paul says there's a law of my mind. If I'm going to set my will to persist, hold my course, get past resistance, and to succeed by getting to where God wants me, I've got to know how the law of my mind works.

If I don't know how the law of my mind works, I could be operating in opposition to that law. If I operate in opposition to the law of my mind *and* the law of God, then I automatically open the door to the law of sin.

There's a law of sin, just like there's the law of gravity, the law of lift, and the law of sowing and reaping. There is a law of sin. If we don't break the cycle of sin in our lives, then the law of sin dictates how things will end. Throughout this book, I've often referred to Galatians 6:9. But let's look at this verse in the context of verses 7 and 8 as well, because it shows the outcome of the law of sin:

> *Do not be deceived, God is not mocked; for whatever a man sows, that he will also reap.* ⁸*For he who sows to his flesh will of the flesh reap corruption, but he who sows to the Spirit will of the Spirit reap everlasting life.* ⁹*And let us not grow weary while doing good, for in due season we shall reap if we do not lose heart.*

Just as there is a law of God and a law of sin, there is a law of our mind. What that means is that we've got to go to work on our minds. A lot of times God wants to do something in and through us and it's our minds that are checking us and preventing us.

If we don't understand the law of our minds and work that law, it will continue to check us and prevent us from moving forward.

In Job 36:11 (NIV) we read, "If they obey and serve him, they will spend the rest of their days in prosperity and their years in contentment." I love this Scripture. This is one of those Scriptures God gave me when I was studying obedience. It says, "If they obey and serve him, they will spend the rest of their days in prosperity and their years in contentment."

Now look at verse 12, "But if they do not listen, they will perish by the sword and die without knowledge." (NIV) Whoa! That's quite a contrast!

Now, there's a bunch of Scriptures like that. For instance, in Deuteronomy chapter 28, God devotes about 14 verses to talking about the blessings that come from following Him. But He spends about 50 verses talking about the curses that will come on the disobedient.

What I'm getting at is this: God is always trying to get us to flow with how He made us. He's always trying to get us to live in concert with how He made us.

If we read the promises of God in Scripture, we'll always see the *good* that will come if we obey Him and the *bad* that will come if we don't because God has designed us and made us to move away from *pain* and run toward *pleasure*. God made us to move away from pain and run toward pleasure. That's why it says, "I will bless you if you obey, but if you don't obey, you will be cursed, you'll die."

What does this have to do with setting our will? This is the law of our minds.

Think about a time when you inadvertently grabbed something that was very hot. When we realize that we've grabbed something hot, we don't just continue to hold on to it. We drop that hot item and move away from the pain. We move away from the pain, because we don't like pain.

The law of our mind says that's the way God made us—to move away from *pain* and to reach toward *pleasure*. The Scripture teaches us that if we understand that God is a good and awesome God, then we realize that serving Him is worth it. Serving Him brings us pleasure. In Deuteronomy 28, God said, "If you obey, the blessing will overtake you. But if you don't obey, all these curses will come on you."

So, if we can associate enough pleasure with obeying God, that pleasure will strengthen us to endure all the temporary discomfort. If we can associate enough pleasure with pleasing God and doing what's right, that pleasure will strengthen us to get past the pain of resistance, hard times, the fact that it looks like it's not working, and other folks' responses. This is the law of the mind at work.

Perhaps on your job, they told you they were going to promote you to manager, but that you needed to prove yourself. So you endure working long hours, you put up with going in to work on the weekend. Why do you do this? Because the company put something out there in front of you and you're willing to endure the hardship in the hope of receiving the promotion. You focus on the pleasure part and persist through the hardships (the pain). This is how God made us.

Jesus focused on pleasure rather than pain when He went to the cross to die for our sins. Hebrews 12:2 says, "Who for the joy that was set before Him endured the cross." The joy and pleasure were out there in the future, not here and now. He endured the pain now for the joy of what lay ahead.

If we can associate enough pleasure with pleasing God; if we can associate enough pleasure with obeying and trusting God, then we can endure whatever hardships and trials come our way right now. This principle or law of the mind is all through the Bible.

But the law of our minds not only applies in the manner we've been discussing, but also in reverse. What I mean by that is that we may be in some destructive behaviors right now. Perhaps we're addicted to some habit, drugs, alcohol, or other destructive behavior. Here's where Galatians 6:7-8 applies again:

> *Do not be deceived, God is not mocked; for whatever a man sows, that he will also reap. ⁸ For he who sows to his flesh will of the flesh reap corruption, but he who sows to the Spirit will of the Spirit reap everlasting life.*

The pleasures associated with these destructive behaviors are temporal and fleshly. And these destructive behaviors all bear serious consequences that destroy our bodies, our minds, our relationships and our lives. We rid our lives of destructive behaviors by seeing and imagining the painful consequences of them and the more repulsive an image we can visualize, the better.

I've never smoked because it was repulsive to me as well. Cigarettes damage every organ in the body and will kill a person. Smoking can shorten one's life. Smoking can cause one's insurance to go up, or even prevent a person from getting insurance. Cigarettes make one smell bad and the second-hand smoke will damage one's children—even worse than the first-hand smoke.

I applied this law of the mind to my life when I considered smoking. I thought about the fact that if I were to smoke, it could cut my life short. Then, somebody else will sleep in the bed I bought. Somebody else will sleep with the wife I had. I imagined the pain associated with smoking and all those thoughts were repulsive to me!

We've got to associate pain with poor choices, because we want to move away from pain. Alcoholism, drugs, pornography—all that stuff we want to move away from because it's destructive.

The Parable of the Prodigal Son is a classic example of what I'm talking about. Remember, the younger son said, "Dad, give me what's coming to me. I don't want to be here. I want to be out on my own." So his father gave him his inheritance and he left home. He went off to a far land with all his money and made life one big party. (See Luke 15:11-32.)

This young man thought that the *pleasure* of being on his own would give him what he wanted in life. He thought it was *painful* for him to stay under his daddy's roof. He said, "I've got to get out of here!" All he could think about were the pleasures of being on his own, having his own money, and doing whatever he pleased.

Something happened though. The Bible says when he ran out of money, his friends left, his girl left, his dog left, everybody left. Then he was forced to go to work on a job way beneath his dignity feeding pigs. He had no money and no food. He was so hungry that he would have eaten the pig slop that he was feeding the pigs, but no one would give him any. They wouldn't even give him pig slop!

At that point, something happened to him. For the first time he realized the *pain* involved in the choices he had made. Now he felt the *pain* of living without support. He felt the *pain* of being abandoned by his friends and feeling like he'd been used. All of a sudden the *pain* of being under his daddy's roof wasn't so painful because, now he had a different reference point for pain.

So, this young man started thinking, "You know what, all my father's servants are eating better than I am." What was this young man doing? He found a new reference point for pain and pleasure. The Bible says, "He came to himself." (Luke 15:17)

So this young man said, "I know what I'll do. I'm going to go back to my father and ask him if he would just let me be a servant in his household. I'm not asking to be restored as his son. I just want to get back into his house. Watch this: he associated *pleasure* with being back under his father's roof and under his care, eating plenty versus the *pain* of not having anything.

God designed us to move away from pain toward pleasure. Here's a key, here's a secret. What keeps me on course is that pain—pleasure reference. Because the pleasure—both present and future—far outweighs the pain.

This is what keeps us consistent. This is what helps us endure. When we set our will, there isn't anybody who can disturb our peace and take away our pleasure. Nobody can take that from us!

I mentioned earlier, the only thing you can control is you. I can't control what other folks do, but I determine what I'll do, I set my will to finish it. It would be very painful for me to break fellowship with God. It is more important for me to have peace with God than to have peace with you. I love you too, but it's even more painful for me to break fellowship with God. My relationship with God is more important than anything else.

That's what Joseph said when Potiphar's wife tried to seduce him. He told her, "Listen, God's favor is on me. When your husband is gone, I'm the man of the house. Your husband put me in charge of everything. It's not about you Mrs. Potiphar. I can't sin against God!" (See Genesis 39.) Joseph put his relationship with God above others and above his natural fleshly desires.

If we can count the cost and if we can set our will to endure temporary pain in order to obtain eternal pleasure, then it becomes a lot easier to persevere and never quit.

We've got to associate pain with disobeying God. We've got to associate pain with shaming God. We have to feel the pain of claiming to be a child of God and living like the devil.

We must feel *God's* pain when we say that we believe Him and do things that make Him weep.

We've got to draw that pain—pleasure reference. It's *painful* not to obey God! In Hebrews 11:24-26 we read:

By faith Moses, when he became of age, refused to be called the son of Pharaoh's daughter, ^{25}choosing rather to suffer affliction with the people of God than to enjoy the passing pleasures of sin, ^{26}esteeming the reproach of Christ greater riches than the treasures in Egypt; for he looked to the reward.

So, I urge you one last time: don't quit! Never give up! Persevere! Endure! Never Quit!

EDITOR'S NOTE

George kindly allowed me to close his autobiography with the following short note.

First, I am grateful to George for giving me the privilege of serving him as ghostwriter for his autobiography. Although our paths had crossed in recent years, I had not known George until we met in June of 2013 and he asked me to work with him on his book.

Getting to know George through this process has been a wonderful experience for me. For the past several years, we've spent many hours together in person, on the phone, texting and emailing. Though as a reader you may not know either of us, I have attempted to write George's autobiography through his "voice." My goal was to ensure that the reader genuinely comes away from this book knowing George's dynamic personality, energy and character and that I remain invisible.

George's remarkable story does warrant telling! I must admit, however, that ghostwriting on behalf of someone else's life is a weighty trust. Yet, writing for George has also been very interesting and fun. This work required a significant amount of research beyond what George could provide me. He was also a delight to work with—a real gentleman and a genuine follower of Christ. With George, what you see is what you get.

I'm also grateful to George's family members and friends of who graciously agreed to let me interview them. Their insights into George's life have added much to his story in content, insight and credibility.

George, thank you for the honor of writing on your behalf! I pray that God uses this book to glorify Himself and encourage and inspire others to persevere and never quit!

— Rob Fischer,
Author and Leadership Coach

RECOMMENDED READING

Biographies

Bush, Barbara. *Barbara Bush: A Memoir*. New York: Scribner, 1994.

Carnegie, Dale. *How to Win Friends & Influence People*. New York: Simon & Schuster, 2010.

Carson, Clayborne. *The Autobiography of Martin Luther King, Jr.* Unknown: Hachette Book Group, 2001.

Ford, Henry. *My Life and Work: An Autobiography of Henry Ford*. Unknown: Greenbook Publications, 2010.

Friendly, Kenneth L. *Never Quit!* (Audio message). Anchorage, AK: Lighthouse Christian Fellowship, 2008.

Hack, Richard. *Hughes: The Private Diaries, Memos and Letters*. Beverly Hills: Phoenix Books, 2007.

Hill, Napoleon. *Think and Grow Rich*. Meriden, CT: Dover Publications, 2007.

Hillenbrand, Laura. *Unbroken: A World War II Story of Survival, Resilience, and Redemption*. New York: Random House, 2010. (The story of Louis Zamperini)

Keller, Helen. *The Story of My Life*. New York: New American Library, 2002.

Lansing, Alfred. *Endurance: Shackleton's Incredible Voyage to the Antarctic.* New York: Carroll & Graf Publishers, 1999.

Metaxas, Eric. *Amazing Grace: William Wilberforce and the Heroic Campaign to End Slavery.* New York: Harper, 2007.

Metaxas, Eric. *Bonhoeffer: Pastor, Martyr, Prophet, Spy.* Nashville: Thomas Nelson, 2010.

Metaxas, Eric. *Seven Men and The Secrets of Their Greatness.* Nashville: Thomas Nelson, 2013

Pearson, Charles T. *Indomitable Tin Goose: A Biography of Preston Tucker.* Mass Market Paperback, 1988.

Powell, Colin. *It Worked for Me: In Life and Leadership.* New York: Harper Collins, 2012.

Roosevelt, Theodore. *Theodore Roosevelt; an Autobiography.* Public domain.

Sledge, E.B. *With the Old Breed.* New York: Ballantine Books, 2007.

Stross, Randall E. *The Wizard of Menlo Park: How Thomas Alva Edison Invented the Modern World.* New York: Crown Publishers, 2007.

Teresa, Mother and Moore, Thomas. *No Greater Love.* Novato, CA: New World Library, 2001.

Truman, Harry S. *Memoirs by Harry S. Truman: 1945 Year of Decisions.* Unknown, 1999.

Cancer Treatment

Food Matters – documentary video and *Food Matters* book by Mark Bittman

Forks Over Knives – documentary video and *Forks Over Knives* book by Gene Stone, Dr. Caldwell Esselstyn, and Dr. T. Colin Campbell.

The Gerson Miracle – documentary video

See also the Gerson Institute Store, http://store.gerson.org/store/Books/

Made in the USA
Middletown, DE
09 September 2022